Blue Kinships

Springer Nature More Media App

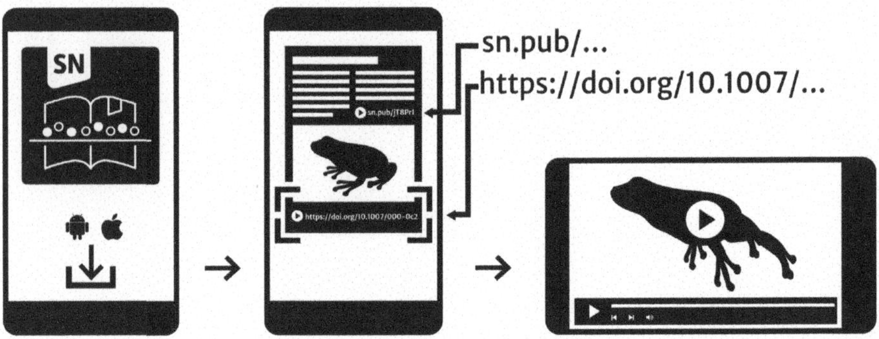

1. Download → 2. Scan Link → 3. Enjoy Video

Support: customerservice@springernature.com

Chiara Certomà
Editor

Blue Kinships

An Exploration of Society & the Ocean

palgrave
macmillan

Editor
Chiara Certomà
Sapienza University of Rome
Rome, Italy

This work contains media enhancements, which are displayed with a "play" icon. Material in the print book can be viewed on a mobile device by downloading the Springer Nature "More Media" app available in the major app stores. The media enhancements in the online version of the work can be accessed directly by authorized users.

ISBN 978-3-031-78618-1 ISBN 978-3-031-78619-8 (eBook)
https://doi.org/10.1007/978-3-031-78619-8

Cover illustration: Cover image graphic by Allegra Guerrazzi

This Palgrave Macmillan imprint is published by the registered company Springer Nature Switzerland AG
The registered company address is: Gewerbestrasse 11, 6330 Cham, Switzerland

Funded by the European Union

New European Bauhaus
beautiful | sustainable | together

Funded by the European Union

#EUmissions
#HorizonEU
#MissionOcean
EU MISSIONS
RESTORE OUR OCEAN & WATERS

Horizontal

New European Bauhaus
beautiful | sustainable | together

Community
New European Bauhaus

Funded by the European Union

The research for and design of this book have been realised by CO>SEA Collaboratorium for Socio-Environmental Analysis of the Ocean, thanks to the Horizon Europe Citizen Science Initiative "SeaPaCS. Participatory Citizen Science against Marine Pollution" funded by IMPETUS4CS project, and the New European Bauhaus project "FishArt. Participatory Art for the Ocean at Anzio's harbour", endorsed by the United Nation Ocean Decade Program as a Decade Action.

This book is a product of the Horizon Europe project "PartArt4OW. Participatory Art for society engagement with Ocean and Water".

To the Ocean.
And those who make it.

Prelude

Tell me about the sea.
Tell me about boats, drawing the shape of the keels with your hands as if to caress them.
Tell me about shrouds, halyards, shrouds…

Picture 1 Ocean making (photo Giuseppe Lupinacci)

Tell me about sails that grip the wind, the rigging that directs their movement, tension, power.
In the wrinkles of the face let me guess the stubbornness of the sun beyond the leech.

And of waves. Tell me about the waves!
Of their impetuosity in the sea turbulence with shining eyes and irreverent smile - of how the sea enters you, your skin absorbs its taste and cracks with saltiness.
Of the majestic and fierce waves when they break and the ocean overwhelms you and exhausts you, takes your breath away and blurs the space.

Tell me about the crossings you have made and those still to be made.
Let me imagine the air of the monsoons in the Indian Ocean, the color of a storm, the boredom of a calm…
What shape do the clouds torn by the Trade Winds have?
What is the blue of the sea farthest from the lands you have explored?
And what is the sound of the Roaring Forties beyond the Great Capes?

Tell me about the ropes that flay the hands, the noise of the rigging in the storms….
Tell me about torn sails and timeless days, of mountains of water that lift the bow, of the unknown life that swarms under the blue to the lightless depths….
Tell me about the stench of the ocean, of the tangle of seaweed in the Sargasso Sea, of the swirling currents in the Bay of Biscay, of the thousand routes between the Southern Cross and the North Star.
And dream me about the sea in winter! When the water becomes crystal and the wind cuts like a blade.

I could listen for hours to talk about the sea.
As it comes, like a whirlwind… without searching for words – alone, with a liquid and restless passion. Whisper to me the silences, the infinite pauses, the wearing quiet of the Ocean.
And then its physical exuberance in which everything becomes real. Even better than real.

And let the other hundreds of questions drown in the abysses that the word cannot tell. (Picture 1)

Turin, Italy Chiara Certomà

Contents

Notes on Contributors

Dror Angel is an applied marine ecologist and head of the Laboratory of Applied Marine Biology & Ecology Research at the University of Haifa. He holds a PhD in Biological Oceanography from the City University of New York and has extensive experience in sustainable aquaculture. Dror's recent research addresses microplastics, invasive species, and jellyfish blooms along the Israeli Mediterranean coast, combining field research with citizen science for broader ecological understanding.

Kristin Bergaust is a pioneer of the self-organized early media art scene in Norway. Her feminist and relational perspectives on contemporary conditions are investigated through performative and technological strategies, sometimes fed by cultural history, scientific data or other narratives. She is a professor at the Faculty of Technology, Art and Design in OsloMet, Oslo Metropolitan University in Oslo, Norway. From June 2024 she took on the function as vice-dean of research at the faculty.

Rihards Bražinskis have graduated bachelor studies on philosophy in University of Latvia. Have worked as a culture journalist and radio broadcaster. Considers the history of graffiti and street art of Riga as his karmic object of research thought the years, collecting and photographing the documentations of the deeds. Not an artist or author, but connected with the scene and with practical knowledge at the same time. Enjoys the aspect of informal pedagogy of it all by having a thematic lectures or making excursions in the city. Also a full time DJ for about a decade, thus intertwining daily with different subcultures and local nightlife in general.

Dr Ruth Brennan is a marine social scientist, integration expert and policy advisor to an Irish Member of the European Parliament, specialising in marine environmental governance, transdisciplinary research and the arts-science-policy interface. She is a Senior Research Fellow at Trinity College Dublin. Her research interests include environmental governance with a focus on society-environment relations, socio-cultural values, socioecological systems and social justice. ruth.brennan@tcd.ie

Luigi Ceccaroni leads strategic innovation at Earthwatch, guiding projects in citizen-science impact assessment, data standardization, and ocean literacy. He holds a BSc in Environmental Sciences, an MSc in Information Technology, and a PhD in Artificial Intelligence, as well as an Executive MBA. Luigi's experience spans from healthcare to environmental decision-support systems, reflecting his commitment to advancing science in diverse contexts.

Chiara Certomà is an assistant professor of socio-political geography at Sapienza University of Rome, with over fifteen years of research experience in participatory methods for science and policy making. Engaged in 15 national and international research projects, is an expert in environmental governance and social innovation for sustainability, adopts participatory citizen science and mobilises volunteers in data collection, elaboration and sharing on the biological consequences of marine plastic pollution. Chiara is actively committed in Marine Social Science, visual geography and ocean studies. She authored more than 30 papers and books and led the EU Horizon sub-project "SeaPacs Participatory Citizens Science against Marine Pollution" (awarded the EU Citizen Science Prize 2024—Diversity and Collaboration), the NEB project "FishArt. Participatory Art for environmental sustainability and aesthetic transformation of Anzio Fishermen's Harbour", and she is now leading the EU Horizon project "Participatory Art for society engagement with Ocean and Water (PartArt4OW)". chiara.certoma@uniroma1.it

Bianca Cruz joined Earthwatch in 2024, following her MSc in Tropical Marine Biology from the University of Essex, where she focused on the contribution of microbial communities in salt marshes, blue carbon status. She holds a bachelors in Advance zoology and biotechnology. She works across four Earthwatch projects—ProBleu, CircleUp, Iliad, and MoRe4Nature. Prior to her current role, Bianca designed nature education materials and worked as an educator in India.

Kaushiki Das is a researcher at Earthwatch Europe, contributing to the ProBleu initiative that promotes ocean and water literacy in schools. She earned an MSc in Environmental Policy and Regulation from the London School of Economics and a BA in English, History, and Political Science from CHRIST University in Bengaluru, India. Her experience spans environmental policy and research at the intersection of environment and governance. Kaushiki works across four Earthwatch projects—ProBleu, CircleUp, Iliad, and CROPS.

Marta Dieviņa is an aquatic biologist and science communicator from Latvia, Europe. She is interested in exploring and finding out how through art and storytelling we can do more diverse and collaborative water ecosystem research and conservation. Marta holds a BSc, MSc in aquatic biology from the University of Latvia and currently she is pursuing B.A. in Culture and Art studies at the Latvian Academy of Culture. Simultaneously she has expanded her experience during exchange studies in Finland (The University of Jyväskylä) and Iceland (The University of Akureyri), as well as international internship in Spain (the University of Santiago de Compostela). For the last ten years Marta has been working with a diverse group of people and institutions, including municipalities, school children, teachers, illustrators, environmental NGOs, and together with them have been exploring - what is aquatic life, what does it mean to us and how can we coexist?

Dori Edelist is a marine macroecologist at the University of Haifa, focusing on fisheries, invasive species, and jellyfish populations. With over thirty peer-reviewed publications, he studies human-ocean interactions, seeking to benefit both. Dori's work also includes citizen science initiatives for monitoring jellyfish and connecting marine science with the public.

Embla Ekström is an interdisciplinary freelance researcher with an MSc in Sea & Society from the University of Gothenburg. Her focus area of research lies with the cultural and natural resource management and how human-nature relations are historically shaped and affected by societal changes. To investigate this, she lets herself be inspired by the common pool resource use within small-scale fisheries and how practitioner translates their local ecological knowledge into sustainable transformative practices. She is currently working as a transdisciplinary project assistant at the Swedish institute for the Marine Environment in Gothenburg. embla.ekstrom@gu.se

Federico Fornaro is managing director of the independent news agency Raw-News Visual Production Agency with 18 year + of experience in documentaries and hard news in Europe, Asia, the Middle East and Africa. Expert in video-documentation and photography, including underwater and drone-based. Jointly with UniTo, he produced the photographic exposition "Explorations in the Plastisphere" (displayed in Turin, Milan, Palma de Mallorca and London) and has been commissioned the visual documentation in SeaPaCS lead by UniTo (the video "In search for Plastic" and the photographic exhibition "Under the sea"). He is Italian stable correspondent for AlJazeera and NBC News; and recurrent correspondent for CBS, Euronews, Sky, CCTV, France24, BBC. Contracted with Greenpeace, Unicef, Unesco, National Geographic, DataLine, Colonial Pictures, Reuters, Italian Minister for Foreign Affars, Lebanon Minister of Turism and Avaaz. federico.fornaro@raw-news.net

Ignacio Gianelli is an inter-transdisciplinary researcher specializing in marine socio-ecological systems, with expertise in small-scale fisheries, climate change, and sustainability transformations. His research focuses on creating opportunities and spaces for collective learning and action among diverse actors through Futures Thinking approaches and leveraging art-science collaborations.

Heather Green is an interdisciplinary visual artist, educator and Associate Professor of Book Arts and Printmaking at the Herberger Institute School of Art at Arizona State University. Her research examines more-than-human encounters in the intertidal and ecological narratives of abundance and loss on a small headland in the Gulf of California in Sonora, Mexico. Whether documenting the movements of the tide, interviewing fishermen, collaborating with scientists, or combing the shoreline—she aims to evoke a nuanced exactitude of place, attuning to its more subtle reaches through a practice of walking and mapping. In order to offer the experience as a multidimensional, multi-sensory immersion, her installations use a wide range of media and interactive elements. Whether turning a handle to view a moving image or taking away a hand-printed card, she invites the audience to engage visually as well as haptically, allowing an opportunity for participation that calls for more sustained attention. heather.j.green@asu.edu

Marien Helmus began his academic career with a BSc in Bèta-Gamma, majoring in biology, at the Universiteit van Amsterdam (UvA). During his undergraduate studies, he gained his first exposure to interdisciplinary and multidisciplinary studies and found his passion in biology and nature. After obtaining his bachelor's degree, Marien pursued an MSc in Freshwater and Marine Biology, also at the Universiteit van Amsterdam. In the final phase of his MSc, he chose a specialization in Science in Society at the Vrije Universiteit Amsterdam (VU). Through this specialization, he combined his passion for marine life with interdisciplinary studies. After completing his MSc, he entered a traineeship focused on sustainability and is currently a strategic member of the nitrogen supervision team at Omgevingsdienst Noord-Holland Noord (ODNHN).

Philipp Hummer since his childhood has been fascinated by nature, especially by entomology and ecology. Being also a citizen scientist since then, he brings a perspective on the practice of citizen participation to the field. With his professional background and know-how of design coming from long years of experience in digital and analog arts, he founded the SPOTTERON App platform together with his colleague Rainer Holzapfel. SPOTTERON is supporting citizen science, monitoring, and public engagement projects across the globe with community-driven and interactive applications and toolkits with high usability both on mobile devices and desktops. philipp@spotteron.net

Stephen Hurrel is a contemporary visual artist with an interest in coastal environments and marine ecologies. He engages with people and place to produce context-based artworks utilising video, photography, sound, text and digital media. Exhibitions include the Venice Biennale and Colombo Biennale, Sri Lanka and he has several permanent public artworks in the UK, Ireland and Tasmania. He studied Fine Art at Glasgow School of Art, has been a Lecturer in UK institutes and Visiting Artist in Norway, Germany, Lapland and Australia. www.hurrelvisualarts.com

Owain Jones is an emeritus professor of Environmental Humanities at Bath Spa University (UK). Trained as a cultural geographer, he became the first Professor of Environmental Humanities in the UK in

2014, and was deputy director of the Research Centre for the Environmental Humanities at Bath Spa 2016–2019. He has published over 90 scholarly articles on various aspects of nature-society relations, and five co-authored/edited books.

Arne Kinds holds a PhD in Economics from the University of Western Brittany. He is a postdoctoral researcher with the EqualSea lab at the University of Santiago de Compostela, Spain. His research examines ownership structures and market concentration within the fishing industry, particularly focusing on industrial fisheries where corporations headquartered in the Global North operate extensively in the Global South. By integrating quantitative analysis of vessel and quota ownership with an in-depth study of fisheries management institutions, Arne explores the broader socio-political and economic dynamics shaping global fisheries and market organization.

Sophia Kochalski is a social fisheries scientist who strives to better understand complex social-ecological systems in order to contribute to the conservation and sustainable and equitable use of our aquatic resources. Her interdisciplinary background includes a B.A. in European Studies from the University of Passau (Germany) and a M.Sc. in Fishery Science and Aquaculture from the Humboldt-Universität Berlin (Germany). She completed her Ph.D. at the University of Liverpool (UK) on conflict and sustainability of small-scale fisheries. Since then, she has worked on fisheries and human-nature relationships within marine, coastal, and freshwater systems across Europe and South America, employed by the Leibniz Institute of Freshwater Ecology and Inland Fisheries in Berlin, the University of Santiago de Compostela (Spain), and Kiel University (Germany). She also spent two and a half years as a consultant in international cooperation within the fisheries and aquaculture sector, and co-led the 'From Sea to Street' project, which focused on marine-themed murals. kochalski@ceos.uni-kiel.de

Natali Lazzari is an interdisciplinary scientist specializing in marine conservation and social-ecological interactions. She earned her Ph.D. in Biodiversity Conservation from the University of Barcelona, where her research focused on the integrated management of coastal marine social-ecological systems. Currently, she is a MSCA Postdoctoral Researcher at the University of Santiago de Compostela and James Cook University, investigating the diverse values of nature in marine environments, particularly those arising from human relationships with these ecosystems. Additionally, she is a co-founder of the Reef Life Survey Europe association, which serves as the European branch of the Australian citizen science program dedicated to collecting standardized underwater biodiversity data. With her research, Natali is committed to promoting sustainable and equitable practices that protect marine biodiversity while supporting the communities that rely on these vital ecosystems.

Gennadi Lessin is a marine systems modeller at Plymouth Marine Laboratory (PML), where he leads the development and application of advanced modelling tools to understand complex biogeochemical and ecological dynamics in marine environments. His research synthesises insights across benthic-pelagic interactions, greenhouse gas dynamics and marine nitrogen cycling. He is one of the lead developers of the modular biogeochemical-ecosystem model ERSEM. Gennadi leads PML's Ocean Literacy impact area, where he builds and strengthens partnerships between scientific research and educational organisations to enhance public understanding of marine science.

Giuseppe Lupinacci is a water photographer. Passionate of photography since the age of 15, he embarked as sailor and then as skipper, from 1999 to 2010, where he improved his artistic skills working in the Maldives, Turkey, Egypt, Antigua, Barbados, Martinique, Sint Maarten, Saint Lucia, Tenerife and sailing three times the Atlantic and the Indian Ocean. Since 2013 he has been working as a freelance aquatic and underwater photographer for media agencies, press, sports institutions and international projects. He recently started a tight collaboration with international socio-environmental project exploring the transformation of oceanic ecosystems and the relationship between society and the ocean (e.g. the EU projects "SeaPaCS. Participatory Citizens Science against Marine Pollution", "FishArt. Participatory Art for environmental sustainability and aesthetic transformation of Anzio Fishermen's Harbour", "Tentacular Thinking") and his photography solo exhibitions has been presented research and institutional venue in Milan, Rome, Palma de Mallorca and London.

Maria Vittoria Marra has a background in marine biology and a PhD in Zoology. In Italy, her home country, she was involved in ocean literacy initiatives carried out with WWF and other national and local organisations. Since she moved to Galway (Ireland) in 2013, she has collaborated with citizen science and science communication organisations, such as Seasearch Ireland and Pint of Science respectively, and has been a member of the Irish Ocean Literacy Network (IOLN) since 2020. In 2021, she became Education and Public Engagement Officer at Galway Atlantaquaria, taking on the IOLN's Secretariat as part of this role until the end of 2024. maria@nationalaquarium.ie

Sean McGrath is a software engineer and a keen amateur musician/songwriter based in Galway, Ireland.

Emma McKinley is a Senior Research Fellow at Cardiff University. Her research focuses on understanding the complex relationships between society and the sea, taking account of diverse perceptions, attitudes and values held by different communities and audiences, and considers how this insight can be used to support effective ocean governance. Recent projects include the SMMR project, Integrating Diverse Values into UK Marine Management, leading work on ocean literacy. Emma is the Academic Lead of the Severn Estuary Partnership, and is the founder of the Marine Social Science Network, a global, interdisciplinary community of marine social science researchers and practitioners. She sits on the UK's National Decade Committee for the UN Ocean Decade, the International Science Advisory Group for MEOPAR and the IOC-UNESCO's Global Group of Expert on Ocean Literacy. mckinleye1@cardiff.ac.uk

Ludovica Montecchio is an interdisciplinary researcher from Italy. She holds a BA in cultural mediation, where she specialized in English and French linguistics, and an MA in environmental humanities. She graduated from her MA with a dissertation titled "Grief and the blue humanities: an oceanic exploration through the metaphor of Medusa". Her research areas span from environmental and cultural anthropology to ArtScience and Ocean Literacy. She just started her PhD within the MEDiverSEAty program, where she will be conducting an ethnography on the Ocean-Archive platform to look into regenerative practices within biological conservation online as well as offline.

Gabriella Palermo is a Postdoctoral Research Fellow at the University of Palermo, where she also obtained her PhD in Geography in 2023. Her thesis Facts and Fabulations. More-than-wet Geographies of the Black Mediterranean received the Award New Geographies and was published by the Italian Publishing House Franco Angeli (2024). Her research interests are focused on Critical Ocean Geographies, Feminist and Postcolonial

Studies, Blue Humanities and Cultural Geography. gabriella.palermo@unipa.it

Stephen Parkinson is a Researcher at Earthwatch and PhD student in the EPSRC Horizon Centre for Doctoral Training at the University of Nottingham. He works across several Horizon Europe research projects (CircleUp, more4nature, ProBleu, Iliad) looking at the "science of citizen science" and trying to improve the way projects involve local communities and members of the public. Previously, Stephen worked in the freshwater team at Earthwatch, supporting research on the FreshWater Watch project.

Noel Pérez was born near the north atlantic ocean, in a fishing village at the northwestern of Spain. Allways he used drawing like a method of navigation to understand and describe internal universes and dump them to the outside. Since 2009 he had worked in the fields of graphic image and mural painting for individuals, companies and institutions, developing a pictorial line inspired by sequential narrative and popular art and making interventions in all kinds of public spaces and for different audiences. His artistic work ranges from illustration to design, and from scenography to large-scale murals in public spaces. The work of Noel can be seen in cities in Spain, Germany, Mexico and the USA. He received scholarships for artistic creation in Cordoba (Spain), Angouleme (France) and Guadalajara (Mexico). In the latter, he completed his training at the University Centre of Art, Architecture and Design and did his first solo exhibition with Gallerie Apollinaire on André Bretón´s cultural center in 2016.

He worked in mural festivals in Mexico (Central de Abasto de la Ciudad de México) and Spain (Rexenera Fest, Vigo Cidade de Cor). He led the Viladomar artistic platform in his hometown where the "Mais alá dos muros" festival takes place. Hola@novenoel.com

Baiba Prūse is an interdisciplinary researcher from Latvia with a focus on local ecological knowledge (ethnobiology) and citizen science. She hold a B.Sc. from Business Academy Aarhus (Denmark) in Environmental management field, M.Sc. from Uppsala University (Sweden) in Sustainable Development and completed her PhD on knowledge co-creation at University of Latvia. She is strongly supporting the research strategy of inter/transdisciplinarity and collaborative work between artists. She is eager to learn more on how to move from research recommendations to direct implementation and foresee—art and science combination as the perfect starting point. In October 2024, Baiba joined UCC MaREI team where with the colleagues she explores citizen engagement across Mission Ocean activities.

Svea Rogge is a B.Sc. student in Ecology and Environmental Planning at the Technical University in Berlin, Germany. Within an internship at the university she worked on narrative analysis and narrative's potential for species conservation. She is keen to work in interdisciplinary and mixed method approaches in the fields of science, arts, linguistics and conservation social sciences. Besides her studies, she works on different environmental topics as a writer and editor in online and print media and supports scientific projects. svea.rogge@campus.tu-berlin.de

Katina Roumbedakis is a MSCA Postdoctoral Researcher at the Instituto de Investigations Marinas (IIM-CSIC, Spain) and Centro de Investigación y de Estudios Avanzados del Instituto Politécninco Nacional (Cinvestav-Mérida, Mexico), developing the project SUCCEPHS: A Pathway Towards Sustainable Use and Conservation of Cephalopods. She holds a PhD and M.Sc in Aquaculture from the Federal University of Santa Catarina (Brazil). Her research has increasingly been interdisciplinary, focusing on small-scale fisheries, sustainability and social equity. During her career, Katina has collaborated with several researchers worldwide, conducting research in Latin America, Europe and Africa. Katina is a culturally sensitive professional with the ability to build meaningful relationships with people from diverse backgrounds.

Estefanía Sánchez-Flores is environmental Engineer from the Universidad de las Americas-Ecuador (2010), with a Master of Science (M.Sc.) in Conservation Biology from the Pontificia Universidad Católica del Ecuador (2016). She is currently pursuing a PhD in Natural Resources and Sustainable Management at the Universidad de Córdoba, Spain, and is a WWF Russell E. Train Education for Nature (EFN) Fellow. Since 2014, her research has focused on conservation and environmental education, utilizing art as an educational and awareness-raising tool. She is a Director of Education for SCB-LACA and is President of Corporación Yemanyá- Agua y Conservación.

Milena Arias Schreiber is a senior researcher in environmental social sciences at the School of Global Studies at the University of Gothenburg (Sweden) and holds currently a Marie Skłodowska-Curie fellowship at the University of Santiago de Compostela in Spain. She has a master's degree in Aquatic Tropical Ecology and a PHD in Social and Economic sciences. Her research focuses on the analysis of human-nature relations using fisheries as a case study and in particular institutional drivers of sustainable fisheries governance. She is doing research on the potential of transforming seafood systems for the revalorization of small-scale fisheries in Sweden. In Spain, she is working on the contribution of Galician women shellfish gathers to coastal communities development and wellbeing. milena.schreiber@gu.se

Cika Schulz is an utrecht-based artist and educator. She gratuated in Fine Arts and Design in Education in 2022 and worked with several partners, among which Street Art Museum Amsterdam. She created several murals in the past three years, using our natural environment as an inspiration, but always reacting to the physical and social environment of the place a mural is set. She also initiated several art projects, including 'Exit Expo' exhibition in a demolition building, empowering both the neighbourhood and young artists from Utrecht.

James Sprinks is a Senior Researcher at Earthwatch Europe, where he focuses on the design and usability of citizen science platforms that ensure both data quality and volunteer engagement. He holds a PhD in Human-Computer Interaction (HCI) and citizen science from the University of Nottingham, as well as degrees in Physics with Astronomy and Geographic Information Systems. Before joining Earthwatch, James was involved in European projects on remote sensing and planetary science, coordinating citizen science platforms like Mars in Motion. He is currently involved in several Horizon Europe projects (CHEERS, ProBleu, RURALITIES, BENCHMARKS), and is PI of the CROPS project tasked with upscaling citizen science.

Enrico Squarcina is full professor and teaches Geography and Didactics of Geography and Cultural Geography at the University of Milano Bicocca. His research interests concern the didactics of geography, political geography, marine environmental education and the cultural and human geography of the sea. In particular, he is dedicated to the study of the cognitive and affective appropriation of pelagic spaces by humans. He is president of the Milan section of the Italian Association of Geography Teachers. He founded GREAM (marine environmental education research group) and was a member of the scientific committee of OLI (Ocean Literacy Italia). enrico.squarcina@unimib.it

Anna Stolyarova is the founder of the Street Art Museum Amsterdam (SAMA) and has over 15 years of experience in leading projects that integrate innovation, creativity, and technology. Through her work at SAMA, she has successfully raised over €500,000 in funding from governmental and private sources, while advising city management and private sector leaders on impactful strategies. As a certified Prince2 practitioner, Anna specializes in digital and e-commerce initiatives, excelling in project management, stakeholder engagement, and creative problem-solving. She is committed to advancing art through technology, empowering end-users, and fostering a global network of art and education pioneers. Her passion lies in building sustainable relationships and delivering user-centered, high-visibility digital solutions.

Tanja Straka is a guest professor in Urban Ecology at the Freie Universität Berlin. Initially trained in urban and bat ecology, she transitioned into conservation social sciences during her PhD, focusing on the cognitive and emotional connections people have with nature. Tanja's research explores among other topics how storytelling, photography, and images shape these relationships. This mural project, with its oceanic theme, marked an exciting new direction in her research, expanding her focus from terrestrial ecosystems to the marine environment. tanja.straka@fu-berlin.de

Geertje Tijsma defended her dissertation entitled 'Embedding Engaged Education; Creating Knowledge together' on the 21st of May 2024. The dissertation discusses strategies for institutional change of universities to connect and bring together different types of knowledge (academic from different disciplines as well as societal or experiential knowledge) in order to address complex societal issues. In her current post doc position at Athena she continues this work specifically looking at creative (artistic) ways to cross boundaries between people and different types of knowledge. g.tijsma@vu.nl

Eduardo Urias is Assistant Professor of Innovation and Participation in Sustainability Transitions at the Athena Institute, VU Amsterdam. His research focuses on inter- and transdisciplinary collaboration and knowledge integration, and how these processes can enable more inclusive, just and equitable sustainability transitions. eduardo.munizpereiraurias@vu.nl

Sebastian Villasante (PhD in Economics) has more than a decade of interdisciplinary work experience in coastal systems, ocean sustainability and equity from local to global. He is a Distinguished Research Professor at the Department of Applied Economics at the University of Santiago de Compostela (USC), and Karl-Göran Mäler scholar at the Beijer Institute of Ecological Economics. He was awarded as a Highly Cited Researcher by Stanford University (2022 and 2024). He is the Director and Scientific Coordinator of the EQUALSEA Lab funded by the ERC Consolidator Grant, Coordinating Lead Author of the IPBES Transformative Change Global Assessment. sebastian.villasante@usc.es

Rubén Vinueza-Chérrez studied marine biology at the "Pontificia Universidad Catolica del Ecuador" (PUCE). He has worked as a researcher-professor at several universities and research centers across Ecuador, contributing to various articles and books. His research focuses on two areas:i) The impact of climate change on sea turtle nesting along the continental coast of Ecuador ii) Environmental education and fostering pro-environmental behaviors among fishers to reduce the incidental capture of marine megafauna. He is currently pursuing his doctoral studies at the University of Córdoba, Spain, and is the co-founder of "Corporación Yemanyá Agua y Conservación," a non-profit organization dedicated to conserving coastal marine and freshwater ecosystems in Ecuador's coastal region.

Anna Witter joined Earthwatch in 2024, following her MSci in Zoology from the University of Exeter, where she researched chemical contaminants in marine environments. Her work spans several Horizon Europe research projects—ProBleu, CircleUp, CHEERS, and MoRe4Nature. Anna previously conducted research on zero-emission marine propulsion systems, contributing to sustainable design in the marine sector, focusing on environmental impacts and the circular economy.

Sasha Marie Woods is the Director of Science and Policy at Earthwatch Europe. With a strong background in biological sciences, she earned her BSc in Natural Sciences from the University of Birmingham, an MSc from Imperial College London, and a PhD in Medicine from the University of Bristol. Sasha's postdoctoral work includes research at Harvard Medical School and University College London. Now leading Earthwatch's Science and Policy team, she ensures detailed scientific standards in tackling pressing climate and biodiversity issues. Sasha has also served as a Teaching Fellow in Applied Neuroscience at King's College London.

List of Pictures

List of Tables

PART I

Thinking the Ocean

CHAPTER 1

The sea, above all

Chiara Certomà

We all came from the sea once, and therefore anyone wishing to understand life on this planet has to first understand the sea. (Svensson, 2020)

The sea calls. You will discover this. All it does, basically, is this: it calls. It never stops, it gets under your skin, it is upon you, it is you it wants. You can even pretend to ignore it, but it's no use. It will still call you. The sea you are looking at and all the others that you will not see, but will always be there, lying patiently in wait for you, one step beyond your life. You will hear them calling, tirelessly. It happens in this purgatory of sand. It will happen in any paradise, and in any inferno. Without explaining anything, without telling you where, there will always be a sea, which will call you. (Baricco, 1993)

C. Certomà (✉)
Dep. Memotef, Sapienza University of Rome, Rome, Italy
e-mail: chiara.certoma@uniroma1.it

C. Certomà (ed.), *Blue Kinships*,
https://doi.org/10.1007/978-3-031-78619-8_1

1.1 Society and the Ocean: In Search of an Emotional Connection

Ecological transformations affecting the Ocean have become a major concern for worldwide public opinion in recent decades. The alteration of the physical and biological state of the sea by human societies has resulted in numerous consequences, including the unsustainable exploitation of marine resources and sea use, pollution of the water system from various sources, the increase in sea temperature with tropicalization effects, water acidification and loss of oxygen, and an increase in underwater noise. The irreversible consequences of these emergencies are already affecting the balance of global ecosystems that support life on Earth. These emergencies have already been widely denounced by the scientific community: "Humankind's relationship with the Ocean is no longer sustainable; we are affecting Ocean ecosystems and resources on a global scale through our overexploitation of the Ocean for food and energy production, tourism, and transportation, and through land-based activities such as atmospheric emissions and discharge of waste (Halpern et al., 2010). Additionally, the cumulative effects of human uses are further changing the Ocean's properties, destroying habitats (Rogers et al., 2020), altering species distributions (e.g. Poloczanska et al., 2016), food webs (e.g. du Pontavice et al., 2019; Pauly & Zeller, 2016), and Ocean circulation and bio-chemistry (e.g. Doney et al., 2009; Hu et al., 2020; Levin, 2018), thus altering the capacity of the Ocean to provide ecosystem services to humankind, such as climate regulation or food production (e.g. Costanza et al., 2008; Cheung, 2018)" (Laffoley et al., 2021, p. 1517).

In 2021, on the belief that the Oceans "hold the keys to an equitable and sustainable planet," the UN proclaimed the Ocean Decade and called scientists, governments, businesses, and civic associations to act not only to scientifically document the ongoing anthropogenic changes but for to propose new solutions, intended to inspire forward-looking legal and regulatory interventions in marine governance (e.g. the "Blue Growth" initiative launched by the European Commission). Following the acceleration of the blue economy paradigm, the relationship between societies and the Ocean is increasingly presented as connected to our economic dependence on the Ocean (Ertör & Hadjimichael, 2020). Therefore, marine-related research is often interpreted as devoted to efficient planning and management (Eikeset et al., 2018; Soma et al., 2019), including research and experimentation in social agency (Oautters et al., 2017).

The proposed initiatives are mostly framed within a utilitarian conception of the economic and political importance of the Ocean (Gonçalves, in press) and the efficient management of its resources (Eikeset et al., 2018), including extraction practices (notably for food or energy sources), logistic or leisure activities regulation, further than the marine geopolitics and bordering issues. Nevertheless, as Leffoley et al. clearly points out: "Beneath the surface, there have been notable conservation successes (e.g. Duarte et al., 2020; Knowlton, 2020), but overall, despite the grand pronouncements, it is fair to say we continue to increase extraction of resources and add more pollutants, discounting the damage to the Ocean as minor problems or acceptable losses, against perceived 'essential' exploitative gain. [...] Despite the Ocean being considered 'remote' and largely inaccessible, this has not protected it from many impacts of human activities" (Laffoley et al., 2021, p. 1314–5).

Beyond falling short of our expectations, adopting a purely utilitarian approach produces two significant gaps in our understanding of the intricate relationships between society and the Ocean. Firstly, a utilitarian reading lets the Ocean fade into the background, serving merely as a silent canvas for economic ventures and geopolitical power plays. As Ellen Marie Krefting (2021) notes, this way "the long-term dynamics of knowledge and conflict of interests that are at the heart of the precarious state of the maritime world are neglected" (p. 1) therefore ignoring the historical formation of the realities of the current ecological crises. Such a perspective neglects the evidence that reshaping humanity's connection with the sea demands transcending its mere economic values.

This leads us to the second gap—the absence of an emotional, affective, and care-based connection between society and the Oceans (Parrain, 2010), a disregard of the Ocean's social and cultural significance beyond its economic value (Squarcina, 2015). To strengthen the bond between society and the Ocean, a sole reliance on technological solutions is insufficient. Presently, Ocean Education endeavors to foster a collective and comprehensive understanding of the reciprocal impact between the Ocean and humanity (EC, 2020; UNDP, 2021). This involves cultivating an emotional perspective and nurturing a caring attitude toward the Ocean (Squarcina & Pecorelli, 2017), with the overarching goal of promoting sustainable behaviors (Drisch, 2015; Jefferson et al., 2015). By adopting an emotional gaze, we may turn the seemingly vast and indistinct space of the Ocean into a meaningful place for everyone; and relational values connecting people and the marine world can trigger behavioral change (McKinley et al., 2023, p. 86).

Following the above-mentioned investigative line, this book delves into the "socio-cultural territorialization" (Squarcina, 2015) of the Ocean and contributes to the emergent movement of ideas and practices that understand the Ocean as a conceptual and physical space for reconsidering our relationship with the complex, heterogeneous, and mutable ecological systems of the Anthropocene. In the following pages, we explore various forms of attachment between human, non-human, and more-than-human societies with the sea, with a specific—despite not exclusive—attention to the geography of the sea (Germond & Germond-Duret, 2017) and to the transdisciplinary marine social science research (McKinley et al., 2020). The overall aim of all our contributions is to foreground the Ocean—or, as the title of this chapter suggests, to devote attention to the sea above all.

1.2 Exploring the Ocean

For quite a long time, following Deleuze and Guattari's distinction between smooth and striated space (Deleuze & Guattari, 1987), the Ocean was regarded as the smooth space par excellence. The smooth space has no borders, it allows completely free movement in any direction; the striated space is a mapped, gridded, conceptualized, and controlled space in which possibilities of movement are severely restricted. While at first sight, the Ocean might seem to fall into the first category—and a long tradition in political thinking reputes so (see Lehman et al., 2021), nevertheless the Ocean's resistance to the delimitation of boundaries does not mean that the Ocean is actually "smooth," for two main reasons.

First, in the real world, there are no pure smooth or pure striated spaces, rather mixture of them. For instance, Maskit (1998) claims, when drawing the border of a wild place on a map, like the Ocean, from being a smooth one, it becomes striated. When we give a name to a place, smoothness disappears; in fact, a place is always a space located, defined and (in some ways) known. Via the cultural apprehension of the smooth space, we make it viable for those who go all over it. The process is exemplified by the transformation of the Oceans from smooth to striated spaces, as Maskit explains: "The sea, while apparently a smooth space, is actually gridded and plotted, charted and divided. Because of the absence of natural 'landmarks', the sea requires the imposition of a grid from outside. There is no pretence here that the striations of the sea follow natural patterns or express what was already there. This transformation from smooth to striated is necessitated because of the sea's very smoothness: it is the problems of navigation in open water that motivated the development of better maps and means of navigation. [...] But it also allows the possibility for groups like Greenpeace or the Sea Shepherd Society to play against those striations (as well as other sets of striations) and produce a form of movement which does not fit in with the patterns the striations should allow" (Maskit, 1998, p. 273–274).

Second, the marine geography tradition itself evolved on the analysis of the striation of the Ocean, "in terms of either the circulation of goods, people and knowledge, and /or as spaces across which power was projected to distant lands (Peters, 2010, p. 1260–1272). For instance, the definition of Exclusive Economic Zones since the end of the Second World War, eminently represents the geopolitical interests associated with a conception of Ocean as a "territory beyond terra" (Peters, 2018), ready to be appropriated, controlled and exploited for national interests. The Oceanic Turn in social research has been—and still is—largely characterized by a representation of the sea as both a frontier and an infinite natural resource for the "blue-green capitalism" (Helmreich, 2012, p. 26). From this perspective, the Ocean represents the "capital's preferred mythic element" (Connery, 1996, p. 289) where combination of geopolitical, biopolitical, ecological, and ontological dimensions are easily mobilized to feed the nationally limited ways of thinking.

Understanding the Ocean as a territory generated a wealth of transoceanic studies on the intersection between capital and space (Rediker, 1987), on race and travels (Gilroy, 1993) and "configured the Ocean as a historical space of transnational capital, empire, and slavery" (DeLoughrey, 2016, p. 33).

With a few exceptions, these narratives largely produce an imaginary that posits the sea as a stage for human history, a narrative of latent surfaces rather than immersions, because the Ocean is conceptualized as the space for theorizing the materiality of history, but it rarely is as material in itself (DeLoughrey, 2007, p. 22). Only recently multidisciplinary researches devoted attention the Ocean per se (Anderson & Peters, 2014; DeLoughrey, 2016; Peters et al. 2022). These overturned the narrative of the sea as a space for human history to happen (see Mentz, 2009; Peters, 2017; Steinberg, 2017; Anderson & Peters, 2014;

Bentley et al., 2016), and open to explorations of the multiple socio-spatialities that emerge from the Ocean's liquid materiality (Anderson & Peters, 2014; Chen et al., 2013; DeLoughrey 2016; Peters & Steinberg, 2019; Peters et al., 2022). Such a new wave in socio-oceanic studies cultivates alternative policies, narratives, and connections related to water. Notably, flow and fluidity are beyond the legislative and normative control of territorial states (Steinberg, 2001, p. 245) and the attention for the Ocean movement, totally extraneous to humanly conceivable time and space, stimulated the attention of scholars engaged with the material (Whatmore, 2002), the volumetric (Peters & Turner 2018), and the new materialism turn (Alaimo, 2012; Neimanis, 2018). Nevertheless, the Oceanic is a particularly fertile research area even for contrasting theoretical perspectives. The "blue humanities" researchers, for instance, introduces the "tidalectics" approach to overcome the new materialism thinking. It interrogates how humans engage with water, utilizing literary, cultural, historical, and theoretical connections and ecologies (Mentz, 2024). To this end, researchers have to immerse in a hybrid world where sea-surface events (e.g. migration, trade, and battles) are complemented with a view of the Oceanic depths where a multitude of aquatic creatures reside, where light behaves differently, and sound travels four times faster than through air. By conceiving this plurality of worlds and their circulation blue humanities defy the constraints of borders and resist the desire for certainty within static and established epistemological and ontological positions. In this regard, Lehman et al. (2021) advance the idea of turbulence as a leading concept: "It seems, perhaps, that no concept more accurately encapsulates both the impetus and result of this turn to the sea than that of turbulence. The turbulence of the Ocean's materiality has become increasingly apparent due to extreme weather, changes to Ocean life, and shifting Ocean circulation patterns. At the same time, turbulence suggests the difference that the Ocean makes to some of the land-based foundational principles that underpin social analysis, including nature, society, humanity, and space" (Lehman et al., 2021, p. 193). From the marine depth perspective, marine histories itself showcases not only the brutality inherent in capitalism and imperialism, but also their unpredictability and internal contradictions. Building upon these considerations, Lehman et al. (2021) suggest that the movements of the Ocean imply a non-linear concept of time, a temporality that could be cyclical, drifting, accumulating in imperceptible ways, or getting stuck and then swiftly dislodged (King, 2019). Turbulences draw attention "to patterns emergent from chaos and complexity, and moreover to the acts of careful attention that refuse binaries between complete knowledge and the unknowable" (Lehman et al., 2021, p. 195).

1.3 Tentacular Thinking

Investigations around the concepts of turbulences, liquidities, and temporalities of the Ocean (Steinberg & Peters, 2015), the multiple social spatialities that emerge from the Ocean's liquid materiality, the more-than-human kins that occur at the sea (Haraway, 2016), and the multiple forms of societal participation in the exploration of recombinant marine ecologies (Brennan, 2022; Buchan et al., 2023) characterize recent research in Critical Ocean Studies. Pivotal here is the work of geographers such as Philip Steinberg (2001) and Kimberley Peters (2010) and their more-than-wet ontologies (Steinberg & Peters, 2019), in which the appreciation for the Ocean materiality, excessive and extensive, helps us to think-with other politics, other possible futures, new ways of living and dying in the multispecies ecology of the Oceanic.

As DeLoughrey (2016) suggests, the interdisciplinary transition toward Critical Ocean Studies provoked a shift of interest from the mobility across transoceanic surfaces to the submersion and the ontological queries this brings about. Steinberg (2013) himself, after having pioneered the social construction of the Ocean recognizes that, despite the sea is a social (i.e. a human) space and construction, it is not just such because there will always be something unfathomable and elusive in the Ocean deepness. He acknowledges that "Indeed, human encounters with the sea are, of necessity, distanced and partial. The encounter from the shore, from the ship, from the surface, or even from the depths, while laden with affective feelings, captures only a fraction of the sea's complex, four-dimensional materiality. To be certain, the combination of emotional intensity with material distance that characterizes our understanding of the sea has made for some excellent literature. Art, after all, thrives on the distance between affective and cognitive understandings. This tension also happens to have led to some relatively enlightened environmental management practices. But the partial nature of our encounter with the Ocean necessarily creates gaps, as the unrepresentable becomes the unacknowledged and the unacknowledged becomes the unthinkable." (Steinberg, 2013, p. 156). Emotional intensity and material distance that characterize our understanding of the sea also determine the distance between its affective and cognitive understanding; therefore our encounter with the sea is unavoidably biased by a sort of partiality and impossible total understanding. While social sciences has often played on this gap—with the somehow paradoxical results of provoking and accentuating the extraneity of the sea—Stenberg suggests that (still not denying the relevance of human history) "in order to fully appreciate the Ocean as a uniquely fluid and dynamic space we need to develop an epistemology that views the Ocean as continually being reconstituted by a variety of elements: the non-human and the human, the biological and the geophysical, the historic and the contemporary." (Steinberg, 2013, p. 157). In this regard, DeLoughrey's analysis of the work of Caribbean writers and artists that theorized the Ocean in terms of convergence of environment and history clearly shows that "Unlike terrestrial space—where one might memorialize a space into place—the perpetual circulation of Ocean currents means that the sea dissolves phenomenological experience and diffracts the accumulation of narrative." (DeLoughrey, 2016, p. 33).

Submerged environments show distinctive and mutable geographies that emerge from ecological networks produced by the bodily interaction of marine creatures with earthly forces and socio-cultural artifacts in the fluid environment of the sea. Lehman et al. (2021), for instance, claim that "Thinking with marine ecologies therefore encourages meditation on the interconnections and indeterminate links among bodies and planetary processes. Rather than a repository, space, or medium upon and within which human systems of governance overlay and play out, the lively Ocean is inter- and intra-active" (p. 206). Investigation into the non-human marine organisms revealed how "alien" elements produce relationships among them and with the humans, raising questions about the possibility to define what life is and what is not, because the assemblages of non-human and more-than-human are increasingly constitutive of the global Ocean. Moreover "marine life-science requires thinking with the radical indeterminacy of these relations in ways that transform how we understand and categorize life and its divergent potentials." (Lehman et al., 2021, p. 206). In this regard, also the language of material semiotic and actor-network-theory, mobilized by new materialism research to account for the non-human agency in marine context, is only partially adequate. As Lehaman et al. (2021) contest, "while it has described assemblages as impermanent and ephemeral, its language often suggests something more machinic, particularly when nodes and networks are described as static concretizations of moments in time" (p. 208), failing thus to account for transformations and turbulences. It is the intimate and distinctive relationship that allows us to perceive the "Oceanic tentacular organisms to serve as figures of a more plural—and, indeed, turbulent—worlding" (p. 209) so that its recombinant organisms, characterized by the indeterminacy of body boundaries, emerge via differentiation both within and from the environment in the "abyssal intimacy" (Picture 1.1) (p. 212).

Living with turbulences and using a language capable of accounting for them has been attempted by the ecofeminist postmodern philosopher Donna Haraway. She advocates for considering non-human organisms not solely as actors but as collaborators in generating concepts and contributing to theoretical interventions. In thinking about the collective futures prefigured by the Oceanic, Haraway contraposes the Anthropocene (which is still characterized by a strong centrality of the human

Picture 1.1 Abyssal intimacy. Looking for biological formations on marine plastic on a fishing vessel (photo Giuseppe Lupinacci)

presence in reading earthy connections) to the Chuthulucene, whose creatures are the "chthonic ones, the not yet finished, ongoing abyssal and dreadful ones that are generative and destructive" (Haraway 2014). Haraway's introduction of the Chthulucene as an alternative to the Anthropocene implies a world that is not completed or stable but is instead "unfinished" and continually engaged in a "troubled" process of evolution. Marine tentacular organisms are proposed as representations of a more diverse—and, indeed, tumultuous—world. The process that creates new kins between societies and the Ocean is called sympoiesis by Haraway. Being aware of its functioning (and being part of it) allows us to "stay with the problem" (Haraway, 2016) and draw on the compositions of materialities, imaginaries, and forces in the liquid space (Neimanis, 2018), to reconceive our common belongingness to the Ocean (Peters et al., 2022). This perspective makes biology appear as "a process of constant change and transformation, of 'posthuman becoming' and 'biocultural hope'".

1.4 Becoming Ocean

Over the past decade, to deepen our understanding and emotionally analyze the relationships between society and the Ocean, relational and affective perspectives have been integrated with natural and social science research results. This required a reconsideration of various spatio-temporal dimensions and the emergence of more-than-human entanglements, interdependencies, transoceanic circulations, of ruptures, and disconnections as generative moves.

Incorporating scientific, artistic, and activist perspectives is vital in investigating these connections, and the emerging hybrid languages often prove to be the only means capable of capturing the complexities of Oceanic. Through the artistic and creative interrogation, in this book, we try to capture, penetrate, and mold the multiple aspects of the relationships between society and the Ocean.

Emergent artistic and cultural expressions engage in reconfiguring the aesthetics in the Anthropocene as a chief means for advancing our understanding. DeLoughrey, for instance, presents the case of the multispecies Museo Subaquático de Arte in The National Marine Park of Cancun, Mexico. Similar to other submarine sculpture museums, art pieces are constructed of inert materials to encourage multispecies colonization, subject to currents and water changes. It brings visitors' attention to the perpetual transformation and fluidity, making visible the colonization of corals on the surface of the sculpture and this interrogates the boundaries of the living and non-living matter. She comments: "The transformations required by these new sea ontologies relict uncomfortable mergers: violence, digestion and indigestion, a radical displacement from places, figurative and otherwise, that we may call home. While earlier scholarship figured the Ocean as blank space or aqua nullius that, in crossing, catalyzed an often masculine agency, the new work of critical Ocean studies engages sea ontologies, figuring maritime space as a multispecies and embodied place in which the Oceanic contours of the planet, including its submarine creatures, are no longer outside of the history of the human" (DeLoughrey, 2016, p. 42).

Thriving on the power of creative and artistic expression to fill the gap between the emotional intensity and material distance with the Ocean, most of these book authors gathered around a discussion group called "SAP&O_Science + Art + Participation about/around/in/on the Ocean." This is an informal context to share experiences and reflect on how the combination of art/creative production, scientific research, and participatory processes can contribute to establishing an emotional connection; and to interrogate transdisciplinary and innovative methodological approaches fostering the collaborations between arts and natural and social sciences for exploring watery socio-ecologies.

Inspired by our engaged research and action art perspective, we particularly work on the possibility of participatory art and creative processes to create a deeper and wider engagement with the Ocean. Participatory art refers to a range of practices where emphasis is placed on the role of the viewer in the physical or conceptual realization and reception of the artwork. It de-emphasizes the role of the professional artist as the sole creator or author of the artwork, fostering social bonds through communal meaning and activity. This approach is informed by social, political, geographic, economic, and cultural imperatives, resulting in works that are open-ended, interactive, and resistant to closure. Public participation in science, planning, and decision-making has already proved to be crucial for understanding societal problems, knowledge production, and collective decision-making because it enhances the democratic quality of societies by opening debates on current conditions and desired states, forcing us to confront disagreements. Similarly, participatory art can act as a powerful tool to captivate the public's interest and fascination with the sea, channel art's symbolic capital toward constructive social change, and inspire, activate, and reinforce collective feelings. While scientific reasoning, in fact, appeals primarily to rational faculties, art resonates with emotions because when the creative process is shared with ordinary citizens, it prompts cognitive, social, and emotional interaction.

Contributions in this book show how we do not merely use our analytic and creative techniques as devices to translate non-human nature wor(l)ds, but also to serve ourselves as translation devices for the Ocean to speak through us. Such an effort demands, as Haraway suggests, "our best emotional, intellectual, artistic, and political creativity, individually and collectively" (Haraway, 2016, p. 208).

1.5 Contents Overview

As described above, for a long time the Ocean has been regarded as the background where historical relationships (economic, social, political, military, and cultural) and trade of people, commodities and capital happened, sidelining its geographical essence. This prevalent view relegated the Ocean to a mere blank canvas where human agency is exerted and resulted in its under-theorization. An opposite tendency overtheorizes the Ocean as a metaphor for challenging modernist notions of identity and subjectivity, unraveling how these concepts are perpetuated through land-centric divisions and representations of space. Building upon and overcoming these perspectives, in the following chapters, through diverse voices and expressive means, we rather advance a comprehension that is both tides to the material, bodily and historical nature of the Ocean, and to the assemblages of humans, non-humans and more-than-humans who made it. This hopefully brings us to an understanding of the Ocean as a space teeming with possibilities, fostering alternative imaginaries, narratives, and politics. To this end, the book proposes a blend of cross-cutting theoretical explorations, field-based examples, and methodological dissertations that scrutinize the emergent blue kinships against the transformations generated by anthropic activities. The preparatory work of this book has been endorsed as a Decade Activity by the United Nation Ocean Decade program.

The first section delves into theoretical exploration, spanning cultural geography, transdisciplinary Oceanic studies, and post-modern philosophical thinking. It reconsiders the narrative, imaginary expedients, and pragmatic conditions influencing the composition of marine life. The second section introduces field-based experimentations aiming to stimulate critical, embedded, and transformative forms of attachment to the sea. This is achieved through participatory, co-collaborative, and posthumanist methods. The interlude offers a visual exploration through artistic photography, illustrating the myriad connections between human, non-human, and more-than-human entities with the Ocean.

First contributor, Emma McKinley, suggests us to the opportunity to position society as part of the solution to Ocean challenges rather than just the cause. Social science research provides valuable insights into human behaviors toward the marine environment, recognizing society's crucial role in sustainable marine resource management. The field of marine social science has experienced significant growth, and forming networks is recognized as an effective way to address environmental challenges collaboratively. This chapter explores the importance of affinity groups and collaborative networks, drawing from existing communities like MARE and the Marine Social Science Network. It emphasizes the essential role of networking in advancing marine social sciences, enhancing Ocean literacy, promoting community engagement, and fostering a sense of belonging to the sea.

Along this argumentative line, Enrico Squarcina introduces us the Ocean as the imaginative last space of freedom, claiming that the sea is the great absentee of the geographical analysis, especially the high seas and to approach it we require a perspective in which physical elements of the Ocean and human behavior are seen in a single framework and for all the interdependencies that bind them; a point of view which makes the culturally central position fully perceived will take on the sea in the future of humanity. He proposes the idea of a socio-cultural territorialization of the sea, through the assignment of names that evoke weather-marine conditions, non-human biological presences, and symbolic values deriving from the relationship of sailors with it and from the projection of human feelings onto the marine spaces.

Diving into the deepness of the Oceanic turbulences, Gabriella Palermo declines research on the cultural geography of the sea with reference to the more-than-wet Ontologies and the Oceanic Turn. The chapter look at some specific trans-Oceanic figurations and more-than-human kin happening at the Sea in which volumes, liquidities and circular temporalities, as properties of this Oceanic materiality, collaborate with the possibility to stay in contact with a turbulence. It describes how whales, human bodies, coral reef act as companion species and weave imaginaries, practices, collaborations, spaces and times of the trans-Oceanic. The Sea, and particularly the High Sea emerges not only as the space where territorialization takes place through geopolitical control, but also as the space in which practices of belonging, care and becoming-with can be regenerated, incorporated, imagined, and so practiced.

By elaborating on new materialisms work to thinking with the Oceanic and develop new ecologies, Chiara Certomà and Federico Fornaro propose an exploration of hybrid assemblages that are re-signifying the ecological global life-supporting systems in the sea. Building upon the material semiotic and politics of more-than-human assemblages, she suggests that Donna Haraway's speculations on the Chthlucene can help us at reversing the capitalocentric narrative of the Anthropocene and coping with pressing environmental problems, notably Ocean pollution. To this end, she engages with the sympoiesis as the chief process forging new kins between societies and the sea so to "stay with the problem" (Haraway, 2016); and explores the emerging ecologies of the plastiphere (i.e. the new biodiversity patterns that have evolved to live on microplastics in marine environments) as exemplary topoi of the Chthlucene.

The Interlude section present artistic photographs by Giuseppe Lupinacci exploring social attachment to the sea via a visual narrative of the unexpected linkages between the physio-chemical, biological and anthropic components of the Ocean. Shots are produced by adopting an approach "from below and inside the sea," showing the actual making of the artistic creation and the emergence of kinship creation, and introduce us to the second section presenting methods and practices.

Here, Stephen Parkinson and colleagues expand on the role of citizen science to allowing people in society to join the debate about complex modern problems and can be used to improve Ocean literacy. She presents the ProBleu (Promoting Ocean and water literacy in school communities) projects facilitates school activities related to Ocean and water literacy through funding calls, the adaptation of existing educational methods and resources, and the development of new resources—including those related to citizen science—to support, strengthen and sustain school activities. Moreover, the ILIAD (Integrated digital framework for comprehensive maritime data and information services) project further provided multiple sources of data, including data from citizen science, to generate digital twins of the Ocean.

Citizen science approach is again at the center of interest of Baiba Prūse and team that innovatively investigated murals in European cities, to express, inspire, and educate on society-Ocean relationships. The team collected marine-themed murals and applied visual analysis, exploring how they construct sea and human-Ocean relationships, and highlights murals' potential to contribute to a sustainable Ocean future.

By mobilizing an entirely different form of artistic production Maria Vittoria Marra describes the power of music to deliver blue education and public engagement. She present the 'Amhráin na Mara' ('Songs of the Sea') project, in which citizens created an original song, and the evolution of this into 'Galway Bay Is Calling' in summer 2023. Developed in collaboration with Music for Galway and Atlantic Technological University, the project aims to raise awareness of climate change impacts in Galway. Choirs and musicians involved attended Ocean literacy workshops, boosting their understanding of the Ocean's influence and their impact on it.

The integration of science results and artistic production characterized the fourth final chapters. The encounter between artist Stephen Hurrel and marine social scientist Ruth Brennan during an art-science expedition in the Outer Hebrides led to four collaborative projects. Recognizing overlaps in methodology and practice, our initial collaboration prompted a five-year exploration of cultural mapping for understanding emotional and socio-cultural values in marine environments. This chapter delves into the behind-the-scenes aspects of these collaborations, sharing insights from 2011–2015. It explores participatory and creative practices used to reveal the emotional connections of communities to the sea and their surroundings, addressing questions of belonging, entanglements with the more-than-human world, and the nature of 'nature.

The Ocean Connections project, led by artist Kristin Bergaust focuses on the fjords, which merges Ocean and inland realms. Guillermo's mathematical modeling with Lagrangian Descriptor and Evangelos' data contribute to visualizing unseen Ocean processes. Sponsored by the European Commission's JRC Joint Research Center, the project uses chaos theory-based animation to depict OsloXord's biodiversity threats. The animation, part of the Naturarchy project, explores eco-aesthetics and relational frameworks, and—inspired by Donna Haraway's sympoiesis—the project aims to reveal hidden Ocean dynamics and environmental challenges.

Permaning on marginal coastal zone, Heather Green and Owain Jones explored the liminal spaces of estuaries, tidal flats, and wetlands in the collaborative project Tidal Timespace. Tidal landscapes, such as estuaries, exhibit ceaseless rhythms and serve as keystone ecosystems with diverse functions. They provide crucial habitats for birds and fisheries, protect coasts, yet face threats from human development and climate change. Palimpsests of these landscapes, captured in plaster casts, reveal unique patterns at low tide. Focusing on Bahía Adair in Sonora, Mexico, and the Severn Estuary in the UK, this chapter employs interdisciplinary methods—arts, sciences, cultural geography, and environmental humanities. Deep mapping, ethnographic interviews, and autotopographical writing are used to weave biocultural specifics, ecologies of memory, and cultures of care into artists' books, tidal sonification, and an immersive art installation. The exploration aims to showcase shared care narratives contributing to a broader process of ecological healing.

Eventually, the final chapter is a visual exploration into society relationship with the ocean from the perspective of visual content professionals, elaborated by Federico Fornaro and Giuseppe Lupinacci.

Overall, the aim of our collection is to offer you a journey in search of an emotional connection between society and the Ocean for tightening "blue kinships" between humans, more-than-humans and the non-living actors that shape the new hybrid marine assemblages.

References

Alaimo, S. (2012). States of suspension: Trans-corporeality at Sea. *Interdisciplinary Studies in Literature and Environment, 19*(3), 476–493.

Anderson, J., & Peters, K. (Eds.). (2014). *Water worlds: Human geographies of the oceans*. Routledge.

Baricco, A. (1993). *Oceano mare*. Rizzoli.

Bentley, J. H., Bridenthal, R., & Wigen, K. (Eds.). (2016). *Seascapes: Maritime histories, littoral cultures, and TransOceanic exchanges*. University of Hawai'i Press.

Brennan, R. (2022). Making space for plural ontologies in fisheries governance: Ireland's disobedient offshore islands. *Maritime Studies, 21*, 35–51.

Buchan, P., Evans, L., Pieraccini, M.,& Barr, S. (2023). Marine citizenship: the right to participate in the transformation of the human-Ocean relationship for sustainability. *PLoS ONE, 18/3*.

Chen, C., MacLeod, J., & Neimanis, A. (Eds.). (2013). *Thinking with water*. Queen's University Press.

Cheng, L., Abraham, J., Hausfather, Z., & Trenberth, K. E. (2019). How fast are the oceans warming? *Science, 363*(6423), 128–129. https://doi.org/10.1126/science.aav7619

Connery, C. L. (1996). The oceanic feeling and the regional imaginary. InW. Dissanayake & R. Wilson (Eds.), *Global/Local: Cultural production and the transnational imaginary*. Duke University Press.

Costanza, R., Pérez-Maqueo, O., Martinez, M. L., Sutton, P., Anderson, S. J., & Mulder, K. (2008). The value of coastal wetlands for hurricane protection. *Ambio, 37*(4), 241–248. https://doi.org/10.1579/0044-7447

Deleuze, G., & Guattari, F. (1987). *A thousand plateaus: Capitalism and schizophrenia*. Athlone.

DeLoughrey, E. M. (2007). Routes and roots: Navigating Caribbean and Pacific Island literatures. University of Hawai'I Press.

DeLoughrey, E. (2016). Submarine futures of the anthropocene. *Comparative Literature, 69*(1), 32–44.

Doney, S. C., Fabry, V. J., Feely, R. A., & Kleypas, J. A. (2009). Ocean acidification: The other CO2 problem. *Annual Review of Marine Science, 1*, 169–192. https://doi.org/10.1146/annurev.marine.010908.163834

Drisch, J. (2015). Territorialisation des mers et des océans: entre mythes et réalités. *Inflexions, 30.*

du Pontavice, H., Gascuel, D., Reygondeau, G., Maureaud, A., & Cheung, W. W. L. (2019). Climate change undermines the global functioning of marine food webs. *Global Change Biology, 26*(3), 1306–1318. https://doi.org/10.1111/gcb.14944

Duarte, C. M., Agusti, S., Barbier, E., Britten, G. L., Castilla, J. C., Gattuso, J.-P. et al. (2020). Rebuilding marine life. *Nature, 580*(7801), 39–51. https://doi.org/10.1038/s41586-020-2146-7

EC. (2020). *Mission starfish 2030*. Publications Office.

Eikeset, A. M., Mazzarella, A. B., Davíðsdóttir, B., Klinger, D. H., Levin, S. A., Rovenskaya, E., & Stenseth, N. C. (2018). What is blue growth? The semantics of 'sustainable development' of marine environments. *Marine Policy, 87*, 177–179.

Ertör, I., & Hadjimichael, M. (2020). Blue degrowth and the politics of the sea: Rethinking the blue economy. *Sustainability Science, 15*, 1–10.

Germond, B., Germond-Duret, C. (2017). Critical geographies of the Ocean: Mobilities, placefulness and maritime relationalism. In J. Monios & G. Wilmsmeier (Eds.), *Maritime mobilities*. Routledge.

Gilroy, P. (1993). *The black Atlantic: Modernity and double consciousness*. Harvard University Press.

Halpern, B. S., Lester, S. E., & McLeod, K. L. (2010). Placing marine protected areas onto the ecosystem-based management seascape. *Proceedings of the national academy of sciences of the United.*

Haraway, D. (2014). Anthropocene, capitalocene, Chthulucene. http://opentranscripts.org/transcript/anthropocene-capitalocene-chthulucene
Haraway, D. (2016). *Staying with the trouble: Making Kin in the Chthulucene*. Duke University Press.
Helmreich, S. (2012). *Alien ocean: Anthropological voyages in a microbial sea*. University of California Press.
Hu, S., Sprintall, J., Guan, C., McPhaden, M.J., Wang, F., Hu, D., & Cai, W. (2020). Deep-reaching acceleration of global mean ocean circulation over the past two decades. *Science Advances, 6*(6), eaax7727. https://doi.org/10.1111/tpj.15043
Jefferson, R., et al. (2015). Understanding audiences: Making public perceptions research matter to marine conservation. *Ocean & Coastal Management, 115*, 61–70.
King. (2019). **The black shoals**. Duke University Press.
Knowlton, N. (2020). Ocean optimism: Moving beyond obituaries in marine conservation. *Annual Review of Marine Science, 13*, 2.1–2.21. https://doi.org/10.1146/annurev-marine-040220-101608
Krefting, E. M. (2021). Maritime modernities: Formats of oceanic knowledge (MaMo), UiO, University of Oslo. https://www.hf.uio.no/ifikk/english/research/projects/maritime-modernities/
Laffoley, D., Baxter, J. M., Amon, D. J., Claudet, J., Hall-Spencer, J. M., Grorud-Colvert, K., Levin, L. A., Reid, P. C., Rogers, A. D., Taylor, M. L., Woodall, L. C., & Andersen, N. F. (2021). Evolving the narrative for protecting a rapidly changing ocean, post-COVID-19. *Aquatic Conservation, 31*(6), 1512–1534.
Lehman, J., Steinberg, P., & Johnson, E. R. (2021). Turbulent waters in three parts. *Theory & Event, 24*(1), 192–219.
Maskit, J. (1998). Something wild? Deleuze and Guattari and the Impossibility of Wilderness. In A. Light & J. M. Smith (Eds.), *Philosophies of place III*. Rowman and Littlefield.
McKinley, E., Acott, K., Yates, K. (2020). Marine social sciences: Looking towards a sustainable future. *Environmental Science & Policy, 108*.
McKinley, E., Burdon, D., & Shellock, R. J. (2023). The evolution of ocean literacy: A new framework for the United Nations ocean decade and beyond. *Marine Pollution Bulletin, 186*, 86.
Mentz, S. (2009). Toward a blue cultural studies. *Literature Compass, 6*(5), 997–1013.
Mentz, S. (2024). *An introduction to the blue humanities*. Routledge.
Neimanis, A. (2018). Queer times and chemical weapons, suspended in the Gotland deep. *Journal of Contemporary Archaeology, 5*(1), 66–78.
Oautters, M., Fernández, P., Pérez, C., & Ferreira, M. (2017). *25 innovative and inspiring solutions to combat plastic marine litter in the Mediterranean Region*.

Regional Activity Centre for Sustainable Consumption and Production (SCP/RAC).
Parrain, C. (2010). *Territorialisation des espaces océaniques hauturiers, Doctoral dissertation*. Université de La Rochelle.
Pauly, D., & Zeller, D. (2016). Catch reconstructions reveal that global marine fisheries catches are higher than reported and declining. *Nature Communications, 7*, 10244. https://doi.org/10.1038/ncomms10244
Peters, K. (2010). Future promises for contemporary social and cultural geographies of the sea. *Geography Compass, 4*(9), 1260–1272.
Peters, K. (2016). *Water worlds: Human geographies of the ocean*. Routledge.
Peters, K. (2017). Oceans and seas: Physical geography. In *The International Encyclopedia of Geography*. Wiley.
Peters, K., et al. (Eds.). (2022). *The Routledge handbook of ocean space*. Taylor & Francis.
Peters, K., & Steinberg, P. (2019). The ocean in excess. Towards a more-than-wet ontology. *Dialogues in Human Geography, 9*(3), 293–307.
Peters, K., & Turner, J. (2018). Unlock the volume: Towards a politics of capacity. *Antipode, 50*(4), 1037–1056.
Poloczanska, E. S., Burrows, M. T., Brown, C. J., García Molinos, J., Halpern, B. S., Hoegh-Guldberg, O. et al. (2016). Responses of marine organisms to climate change across oceans. *Frontiers in Marine Science, 3*, 62. https://doi.org/10.3389/fmars.2016.00062
Rediker, M. (1987). *Between the devil and the deep blue sea: Merchant seamen, pirates, and the Anglo-American maritime world, 1700–1750*. Cambridge University Press.
Rogers, A., Aburto-Oropeza, O., Appeltans, W., Assis, J., Balance, L.T., Cury, P. et al. (2020). Critical habitats and biodiversity: Inventory, thresholds and governance. *Blue Paper 10*. Report prepared for the (Norwegian) Prime Minister's high level panel on a sustainable oceaneconomy. Retrieved from https://www.oceanpanel.org/blue-papers/critical-habitats-and-biodiversity-inventory-thresholds-andgovernance
Soma, K., van den Burg, S. W. K., Selnes, T., & van der Heide, C. M. (2019). Assessing social innovation across offshore sectors in the Dutch North Sea. *Ocean & Coastal Management, 167*, 42–51.
Squarcina, E. (2015). *L'ultimo spazio di libertà. Un approccio umanistico e culturale alla geografia del mare*. Milano.
Squarcina, E., Pecorelli, V. (2017). Ocean citizenship. The time to adopt a useful concept for environmental teaching and citizenship education is now, J-Reading. http://www.j-reading.org/index.php/geography/article/view/168
Steinberg, P. (2001). *The social construction of the ocean*. Cambridge University Press.

Steinberg, P. (2017). Oceans and seas: Human geography. In *The International Encyclopedia of Geography*. Wiley.
Steinberg, P., & Peters, K. (2013). Wet ontologies, fluid spaces: Giving depth to volume through oceanic thinking. *Environment and Planning D: Society and Space, 31*(2), 232–250.
Steinberg, P., & Peters, K. (2019). Wet ontologies, fluid politics: The sea in human geography. Routledge.
Steinberg, P. E. (2013). Of other seas: Metaphors and materialities in maritime regions. *Atlantic Studies, 10*(2), 156–169.
Steinberg, P., & Peters, K. (2015). Wet ontologies, fluid space. *Environment and Planning D, 33*(2), 247–264.
Svensson, P. (2020). The book of eels: Our enduring fascination with the most mysterious creature in the natural world. HarperAudio
UNDP. (2021). One ocean. https://oneocean.undp.org/
Whatmore, S. (2002). *Hybrid geographies: Natures cultures spaces*. SAGE.

CHAPTER 2

Building Relationships and Networking for Marine Social Sciences

Emma McKinley

2.1 Understanding and Mapping the Marine Social Science Community

Given the challenges confronting marine and coastal management, there is a growing push for better integration of humanities and social science research and practice with the more traditional, natural science-orientated approaches (Holm & Brennan, 2018; Holm et al., 2015; McKinley et al., 2022). This integration of social sciences is integral to both effectively understanding the complexities of the problems and finding feasible, sustainable solutions. Moreover, increasing understanding of how people use and respond to marine and coastal environments could also be an opportunity to frame society as part of the solution to these challenges (see for example Vince & Hardesty, 2018), rather than solely being viewed as the cause. Social science methods and concepts provide a valuable lens and a diverse toolbox through which the relationships between different groups of society and the ocean can be viewed and understood,

E. McKinley (✉)
Cardiff University, Wales, UK
e-mail: mckinleye1@cardiff.ac.uk

C. Certomà (ed.), *Blue Kinships*,
https://doi.org/10.1007/978-3-031-78619-8_2

providing much-needed insight into the drivers and motivations behind a whole suite of human behaviours towards our marine environment (Bennett, 2019; Jefferson et al., 2015; McKinley et al., 2020).

Increasingly, there is recognition from international initiatives that society has a very real, and necessary, role to play in supporting sustainable use and management of our marine and coastal resources. This can be seen, for example, within the United Nation's Sustainable Development Goals, but also within the United Nations Decade of Ocean Science for Sustainable Development (2021–2030) (referred to as the UN Ocean Decade, hereafter), its recognition of ocean science as being 'broad', and the delivery of its ten challenges (Vadrot et al., 2022). Challenge 10, specifically, seeks to 'Restore society's relationship with the ocean' and seeks to ensure that '*Meaningful society-ocean connections are strengthened, driving increased motivation, capability and opportunity for people, across all sectors of society, to make decisions, act and behave in ways that ensure a healthy ocean*' (Glithero et al., 2024). Social science concepts and approaches, therefore, are increasingly positioned as mechanisms through which the restoration of human-ocean relationships can be achieved in a way that is beneficial to both ocean health and societal wellbeing (Claudet, 2021; Spalding & McKinley, 2024).

Marine social science is a rapidly developing field of research (Bennett, 2019; Bennett et al., 2017; McKinley et al., 2020, 2022), encompassing a wide range of disciplines and topics. Encompassed within this are topics including but not limited to: ocean literacy (Fauville et al., 2024; McKinley et al., 2023; Shellock et al., 2024), marine citizenship (Buchan et al., 2023; McKinley & Fletcher, 2010, 2012), behaviour change including fisher behaviour (Andrews et al., 2021), social values (Engel et al., 2020), health and wellbeing (Borja et al., 2020; Fleming et al., 2024) anthropology, culture and heritage (Flannery et al., 2022; Ounanian et al., 2021; Trakadas et al., 2019), public perceptions (Jefferson et al., 2015), and public engagement in decision-making (Kelly et al., 2021). In addition to the more traditional interpretations of social sciences, more recently there has been a move towards the consideration and inclusion of arts and humanities as innovative and inspiring mechanisms to cocreate knowledge, raise awareness, engender connection, and educate different audiences on marine issues (Brennan, 2018; Bennett & Roth, 2019). Despite this relatively recent shift, it is worth noting that while the debate on giving more credence to social sciences in relation to the natural world continues, it is by no means a new conversation. In 2003, Mascia and co-authors set out their thoughts on the need for conservation to take greater account of the social sciences in order to properly understand and inform conservation, policy and practice (Mascia, 2003). Over the past 15 years or more, this conversation has continued, has garnered more support, but significant progress has not yet been realised—there remains a need to continue to speak out about the virtues and value of looking not only to the biological sciences to deal with the challenges facing our natural world, and calls to seek insight from social sciences remain (O 'Leary et al., 2021)—as clearly evidenced within the chapters of this book.

Notwithstanding the recent growth in marine social science activity, and crucially a corresponding acknowledgement of the role of this research within global marine and coastal management and decision-making (Bennett, 2019; McKinley et al., 2020; Partelow et al., 2023), persistent challenges remain. These include, but are not limited to, a lack of appropriate funding structures to support the type of inter and transdisciplinary research often suited to marine social science research; limited marine social science research and practitioner capacity; lack of recognition, understanding and visibility of marine social science research concepts, and methodological approaches; challenges in the relationship between marine social sciences and other disciplines; and, a lack of effective pathways for marine social science research to meaningfully influence ocean governance and decision-making (McKinley et al., 2022; Popova et al., 2023; Spalding & McKinley, 2024). While the solutions to address these challenges are multifaceted, and in some instances, require systemic changes, it is of note that establishing inclusive networks, building collaboration, enlisting champions, and promoting marine social sciences have been identified by the marine social science community as enablers and as central to the advancement of the research (McKinley et al., 2022).

2.2 Building Relationships and Networking for Marine Social Sciences Research and Practice

Academic research as historically, on the whole, been undertaken in a disparate and fragmented way, with multiple organisations, researchers, and sectors involved, meaning that communication and collaboration is not always as effective as it perhaps should or could be (Stojanovic et al.,

2009). As global efforts to address the complex challenges facing the environment, collaboration and inter- and transdisciplinary are increasingly being called for, and, indeed, championed by research funders. Establishing networks, therefore, is an increasingly recognised mechanism for bringing together diverse actors, representing different disciplines, worldviews, career stages, and experiences, with more frequent and meaningful engagement, moving away from more traditional siloed ways of working (Cassi et al., 2008; Roome, 2001).

At their very simplest, networks comprise of individuals or organisations that are linked through some form of similar interest or connection (Borgatti & Halgin, 2011). Networks can be singular in their discipline of topic, or they can be formed around multidisciplinary, or indeed, transdisciplinary interests—and while, the purpose of networks can vary, in the context of academia, it is often centred around knowledge sharing and problem solving across a wide range of themes (Dalton et al., 2020; Keast et al., 2004). Formation of networks can not only result in more impactful research; they can also foster social capital between researchers, broaden and develop existing knowledge, enhance understanding and interdisciplinary working, and foster the development of professional relationships and collaborations, strengthening communities of research and practice, particularly those at an early stage (Cassi et al., 2008; Stojanovic et al., 2009). Networks with diverse memberships encourage the establishment of affinity groups, affording members access to a wide range of skills, experiences, worldviews and interest, and provide an opportunity for supportive social learning (Gates et al., 1997; Musgrove et al., 2024). In addition, while working in the environmental sector offers many opportunities to travel and experience parts of the world and nature that others do not get to see, there is no ignoring the fact that working against the backdrop of extreme biodiversity degradation and loss, and the ongoing climate crises can result in, understandable, yet significant ecological grief or ecoanxiety (Cunsolo & Ellis, 2018). The emotional toll on researchers working ecosystems experiencing unprecedented levels of change and threat is becoming increasingly apparent (Conroy, 2019). There is a much-needed role, therefore, for networks and groups to act as a mechanism for building resilience and support within the scientific community as we continue to push for the changes that are so urgently needed (Conroy, 2019; Cooke et al., 2024).

While there have been active communities of marine social scientists for some time, the community has experienced something of a growth spurt in recent years and, as a recognised discipline, it is relatively young. Over the years, numerous authors have set out calls for increased social science contributions to marine and coastal management (see, for example, Kaplan & McKay, 2004; Van Putten et al., 2021). However, it is only relatively recently that these calls, echoed in the Manifesto for Marine Social Sciences (Bavinck & Verrips, 2020) and from other scholars (e.g. Bennett, 2019; McKinley et al., 2020), have led to marine social sciences finally being explicitly recognised as its own research field. Despite being a seemingly young area of research, the marine social science community already boasts a number of networks through which researchers, practitioners and those with a general interest can build connections, relationships, promote research outputs and develop collaborations (see Table 2.1 for a description of just some of these). There are, of course, other networks and groups, who have been working within or alongside marine social science research and practice. These include the National Marine Educators Association (NMEA) and the European Marine Science Educators Association (EMSEA), both with a focus on ocean literacy and marine education; the Marine Alliance for Science and Technology Scotland (MASTS), who have championed interdisciplinary research, including of social sciences, in recent years, among many others. These networks, and others not listed here, have become recognised ambassadors for marine social sciences, and have been central to the development of marine social science research to date. This has been supported through organisation of conferences (e.g. the biannual People and the Sea Conference hosted in Amsterdam by MARE, or the Society and the Sea conference organised by first by the Greenwich Maritime Centre, and now by National Maritime and the University of Greenwich), monthly newsletters (such as that distributed through the Marine Social Science Network), and general promotion and championing of marine social science research and its place within wider ocean science and governance (e.g. through webinars, social media, and publications).

While one key benefit of network membership may be the opportunity promote research outputs, it is the building of relationships, not only research collaborations, but also friendships that the next section of the chapter will explore in more detail. The importance of affinity groups and the feeling of belonging that comes with group membership is known to enhance feelings of self-efficacy and the desire to remain part of a wider community (Musgrove et al., 2024). A sense of belonging can be defined as having feelings of mattering, being cared about and accepted, feeling

Table 2.1 A description of a selection of existing marine social science networks

Network	*Description and goals*
MARE	Established in 2000 by the University of Amsterdam and Netherlands's Institute for Social Science Research, the Centre for Maritime Research (MARE) is an interdisciplinary social science organisation focused on topics relating to coasts and seas. With six institutional partners spread across Europe, MARE hosts a biennial conference (People and the Sea) and publishes the Journal of Maritime Studies
Marine Social Science Network	Launched in September 2018, the Marine Social Sciences Network is an interdisciplinary and international network working to bring together a growing community and facilitate knowledge exchange between diverse stakeholders from across the marine and coastal sector. Increasing awareness and understanding of the integral role that social sciences can have in management and decision-making for our global seas and coastline is key. The Network is building a multidisciplinary community, facilitating collaboration and dialogue across the marine sciences, economics, the arts, and humanities
People Ocean Planet	Formed by the Marine Alliance for Science and Technology for Scotland (MASTS) in 2020, POP aimed to merge marine natural sciences with social sciences with a view to driving lasting behavioural change to address the challenges facing the global ocean and reduce society's impact on the ocean. Priority ocean issues include climate change, sustainable seafood, land-derived pollution, and deep-sea mining

(continued)

Table 2.1 (continued)

Network	*Description and goals*
Ocean Nexus	Established through the Nippon Foundation, the Ocean Nexus is a global community of ocean researchers and practitioners aimed at ensuring social equity is at the centre of ocean governance. Their work emphasises historically marginalised communities, voices and values, and centres intersectional, anti-racist, anti-colonial approaches, covering nine themes of research: Blue Economy; Ocean Governance, Law, and Policy; Sustainable Development Goals; Fisheries and Food Sovereignty; Marine and Coastal Management; Climate Adaptation; Marine Conservation; Ocean Pollution and Well-Being; and Digital and Data Technology for Oceans
Centre of Marine Socioecology	The Centre for Marine Socioecology (CMS) was established to address the current and future use of our marine coasts and oceans. CMS is a unique collaboration between the University of Tasmania and the Commonwealth Scientific and Industrial Research Organisation (CSIRO), with support from the Australian Antarctic Division. The Centre brings together disciplinary expertise in physics, law, economics, biology, sociology, psychology, human health, art, media, philosophy, and governance. CMS uses this expertise to focus on the complex issues that are developing in the management of the marine estate

that you and your viewpoints are respected and appreciated and that you are a valued member of a community (Estrada et al., 2019; Musgrove et al., 2024). Academia, and research more generally, can be isolating—indeed, these feelings of isolation and being separate from other areas of their home organisations were identified by members of the marine social science community, and even highlighted as a rationale for the formation of the Marine Social Science Network, listed above (McKinley et al., 2020). However, as outlined by a recent horizon scan of the marine social science community; it is a geographically disparate community, covering a diverse range of topics and disciplines (McKinley et al., 2022). How then, can feelings of true belonging and mattering be fostered in a community that is spread across borders and time zones? What is the role of networks for the future of marine social sciences research?

2.3 Case Study: The Marine Social Sciences Network

Drawing on the experience of the Marine Social Science Network (hereafter the MarSocSci Network or MarSocSci), the potential benefits of establishing effective communities of practice and networks are evident. Established in September 2018, following a co-development workshop with marine stakeholders in the UK in January of the same year, the Marine Social Sciences Network aims to be an interdisciplinary and international network working to bring together a growing community and facilitate knowledge exchange between diverse stakeholders from across the marine and coastal sector. Over a series of workshops held throughout 2018, the objectives of the MarSocSci Network were established as:

- Provide a 'home' for a growing online community of marine social science researchers, representing a range of disciplines and nations, and feeding into growing areas of research.
- Provide a platform for dialogue, discussion, and collaboration.
- Promote social science as the gateway to embedding wider societal views and values into policy and management and look at mapping social sciences against the global challenges.
- Provide a role for marine social sciences as science and evidence across policy and management processes.
- Position marine social sciences alongside natural and physical sciences, demonstrating the social value of marine systems and used to support communication and articulation of the complex and complicated relationship between society and the global seas and coasts.

Now, with over 1500 people, representing ocean researchers, practitioners, and even policymakers, signed up to the monthly newsletter, MarSocSci has sought to respond to the needs of the marine social science research community since its inception. This has included workshops at relevant conferences (e.g. Society and the Sea—London, MARE People and the Sea—Amsterdam, International Marine Conservation Congress, Malaysia), a webinar series showcasing the application of diverse types of research from across the marine social science community, and, of course, the formation of its monthly book club group.

In the early months of the COVID-19 pandemic in March 2020, as borders began to close and the much-anticipated events of 2020, as would become inevitable, 2021, started to be cancelled, those feelings of isolation and separateness were exacerbated. In a bid to assuage some of these feelings, the Marine Social Sciences Network established its Book club. The Marine Social Sciences Network BookClub, hereafter MarSocSci Book Club, ran its first session in March 2020. As the world adjusted to an unexpected way of living and 'staying home', the MarSocSci BookClub began as a small group of six marine social science researchers, who came together not only for the promise of reading new books, but also for the personal connections that the monthly chats facilitated (Pictures 2.1 and 2.2).

At the time of writing, the MarSocSci BookClub is in its fifth year, with members reading their 54th ocean themed read (displayed in Photo 2)—and the BookClub mailing list now reaches over sixty people. Despite the online working fatigue experienced by many (Weber et al., 2023), MarSocSci BookClub has, thankfully, long outlasted the COVID-19 restrictions on gatherings and cross-border movement. While those in-person interactions are now very much back on schedule for most working

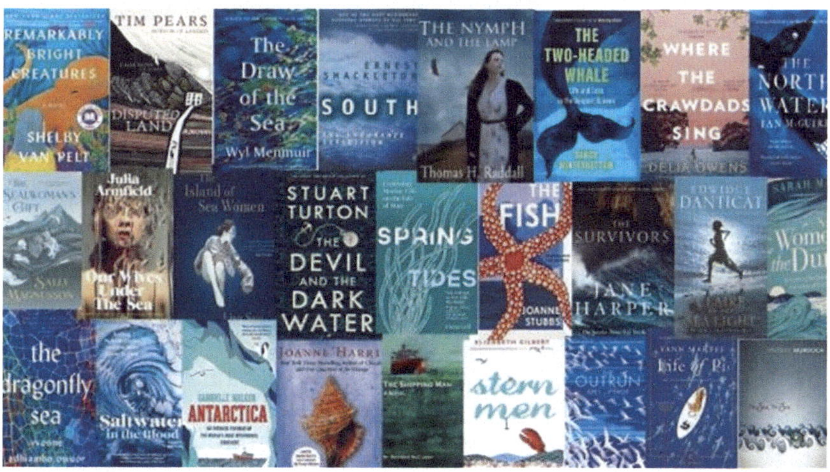

Picture 2.1 A selection of the ocean reads from MarSocSci BookClub (photo Emma McKinley)

Picture 2.2 Poster promoting the MarSocSci BookClub reads for July to September 2024 (photo Emma McKinley)

within academia, the appointment for the monthly online MarSocSci BookClub chats remain firmly in its members' diaries (Picture 2.3).

Understanding the reasons behind the success behind the MarSocSci BookClub requires its own form of marine social science research. Anecdotally, however, initially formed at a time when opportunities for networking and building personal connections (which are so central to academia), the online chat sessions underpinning the MarSocSci BookClub offered a virtual way of maintaining, and initiating, social relationships (Dalton et al., 2020; Gregersen et al., 2023). It gave members an opportunity to virtually leave their homes, their towns, their countries. More broadly, the role that the BookClub has taken on as an affinity group for its members—a safe, respectful, and, crucially, a kind, place for discussion about topics that the group members are all interested in, where friendships and research collaborations can be, and have been,

Picture 2.3 A collage of group photos from the monthly MarSocSci BookClub chat sessions on Zoom (photo Emma McKinley)

fostered. The emphasis on kindness as being central to the BookClub is important to note here. Recent publications have recognised that kindness in research and academia is not always easy—notably, a paper by Boulter et al. (2023) explores the challenges of engendering and maintain kindness in academia and research, considering this through the lenses of being kind to yourself, as well as to colleagues and peers and presents a framework for how to foster kinder teams within science. The definition of kindness is multifaceted and not as simple as we might expect—the dimensions included in this framework include ensuring your passion for your topic is sustained, reduction in stress (something that is perhaps easier said than done), and fostering inclusive, supportive research and collaboration teams (Boulter et al., 2023). The MarSocSci BookClub has fostered a kind environment; one where affinity with each other and with the ocean theme of the BookClub reads, a feeling of you and your views, thoughts and ideas mattering and being supported, of there being no stupid questions and of no judgement (as long as the kindness rule is adhered to!) are at its core.

2.4 Looking to the Future and Concluding Comments

As stated within the UN Ocean Decade implementation plan *'Ocean science is broad. It encompasses natural and social science disciplines, local and indigenous knowledge, it includes the science-policy and science-innovation interfaces...'* (UNESCO-IOC, 2021). With the recognition of this breadth of ocean research, it stands to reason that the role of marine social science research will only continue to deepen, with recurrent calls for integration, interdisciplinary working, and better understanding of human-ocean relationships clearly being increasingly recognised as being not only central to, but fundamental, to broader ocean discourse and addressing the current challenges at the pace and scale required. As this diverse and dynamic community of researchers and practitioners continues to develop, networks will be central to fostering collaboration, minimising duplication of efforts, and maximising the impact of research on effective ocean policy making.

Through this chapter, there have been examples of both long-standing (MARE) and more recent (People Ocean Planet) networks focusing on understanding the human dimensions of the ocean, in a bid to delivering a healthy ocean for the future. It is important to note here that networks are not always successful—often, they are formed with little to no funding behind the, a passion project from their founders (Dalton et al., 2020). Regardless of the level financial backing, a successful network requires significant investment of time—to build relationships, to create platforms that allow for collaboration, to promote the networks, to put together the newsletter—the list is, at times, endless. Despite the challenges outlined earlier in this chapter, it cannot be ignored that their role in advancing marine social sciences research, and indeed championing its role alongside other aspects of ocean science within ocean policy discussions, has, and continues to be, fundamental. As stated by Dalton et al. (2020), networks are often uniquely positioned to catalyse change and can facilitate capacity building, knowledge sharing and exchange, and co-development of solutions that respond to changing needs within the ocean science and policy landscape. The social sciences in the conservation field more broadly continue to receive a poor percentage of funding, and coverage within academic publications, and the natural sciences continue to dominate the global discourse of the interrelated crises of biodiversity loss and climate change (see Overland & Sovacool, 2020; Partelow et al., 2023).

However, calls for change are getting louder, and it is increasingly evident that a robust global network, or indeed connected networks, of marine social scientists can act as a catalyst and facilitator towards the shift that is needed. As explored in this chapter, these networks can (1) foster global collaborations for theoretical insights and advancements in the marine social science field and (2) enhance awareness of the importance of social sciences in marine conservation and planning for (3) increase influence of the marine social sciences in regional, national, and global decision-making processes and platforms. Effective networks offer opportunities for cross pollination of ideas by bringing together different communities of research and practice, drawing expertise and insights from traditional marine social science disciplines, embracing natural and physical science knowledge, but, also, expanding into different areas of scholarship, including the arts and humanities, local and Indigenous knowledge and experiences (Glithero et al., 2024; Jung et al., 2022) and across the science-policy-practice interface. Despite the Marine Social Sciences Network being relatively young at the time of writing, the engagement and growth it has experienced since its launch in 2018 clearly indicate that there is space, and indeed a demand, for networks to bring together and provide a platform to support, and indeed advance, the aspirations, actions, and goals of the global marine social science community.

Crucially, as we strive to address the myriad challenges facing the ocean and, indeed ensure the resilience of our communities' capacity to continue these efforts, there is a need to push against historically siloed ways of working, however established they might be, and to embrace and foster open, inclusive, collaborative, and *kind* communities of research and practice.

References

Andrews, E. J., Wolfe, S., Nayak, P. K., & Armitage, D. (2021). Coastal fishers livelihood behaviors and their psychosocial explanations: Implications for fisheries governance in a changing world. *Frontiers in Marine Science, 8*. https://doi.org/10.3389/fmars.2021.634484

Bavinck, M., & Verrips, J. (2020). Manifesto for the marine social sciences. *Maritime Studies, 19*(2), 121–123.

Bennett, N. J. (2019). Marine social science for the peopled seas. *Coastal Management, 47*(2), 244–252.

Bennett, N. J., Roth, R., Klain, S. C., Chan, K., Christie, P., Clark, D. A., Cullman, G., Curran, D., Durbin, T. J., Epstein, G., & Greenberg, A. (2017). Conservation social science: Understanding and integrating human dimensions to improve conservation. *biological conservation, 205*, 93–108.

Bennett, N. J., & Roth, R. (2019). Realizing the transformative potential of conservation through the social sciences, arts and humanities. *Biological Conservation, 229*, A6–A8.

Borgatti, S. P., & Halgin, D. S. (2011). On network theory. *Organization Science, 22*, 1168–1181. https://doi.org/10.1287/orsc.1100.0641

Borja, A., White, M. P., Berdalet, E., Bock, N., Eatock, C., Kristensen, P., Leonard, A., Lloret, J., Pahl, S., Parga, M., & Prieto, J. V. (2020). Moving toward an agenda on ocean health and human health in Europe. *Frontiers in Marine Science, 7*, 37.

Boulter, J., Orozco Morales, M. L., Principe, N., & Tilsed, C. M. (2023). What is kindness in science and why does it matter? *Immunology and Cell Biology, 101*(2), 97–103.

Brennan, R. E. (2018). Re-storying marine conservation: Integrating art and science to explore and articulate ideas, visions and expressions of marine space. *Ocean & Coastal Management, 162*, 110–126.

Buchan, P. M., Evans, L. S., Pieraccini, M., & Barr, S. (2023). Marine citizenship: The right to participate in the transformation of the human-ocean relationship for sustainability. *PLoS ONE, 18*(3), Article e0280518.

Cassi, L., Corrocher, N., Malerba, F., & Vonortas, N. (2008). Research networks as infrastructure for knowledge diffusion in European regions. *Economics of Innovation and New Technology, 17*(7–8), 663–676.

Claudet, J. (2021). The seven domains of action for a sustainable ocean. *Cell, 184*(6), 1426–1429.

Conroy, G. (2019). Ecological grief grips scientists. *Nature, 573*(7774), 318–319.

Cooke, A., Benham, C., Butt, N., & Dean, J., (2024). Ecological grief literacy: Approaches for responding to environmental loss. *Conservation Letters*, e13018.

Cunsolo, A., & Ellis, N. R. (2018). Ecological grief as a mental health response to climate change-related loss. *Nature Climate Change, 8*(4), 275–281.

Dalton, K., Skrobe, M., Bell, H., Kantner, B., Berndtson, D., Gerhardinger, L. C., & Christie, P. (2020). Marine-Related learning networks: Shifting the paradigm toward collaborative ocean governance. *Frontiers in Marine Science, 7*, Article 595054. https://doi.org/10.3389/fmars.2020.595054

Engel, M., Vaske, J. J., & Bath, A. J. (2020). Value orientations and beliefs contribute to the formation of a marine conservation personal norm. *Journal for Nature Conservation, 55*, Article 125806.

Estrada, M., Young, G. R., Nagy, J., Goldstein, E. J., Ben-Zeev, A., Márquez-Magaña, L., & Eroy-Reveles, A. (2019). The influence of microaffirmations on undergraduate persistence in science career pathways. *CBE—Life Sciences Education, 18*(3), ar40.

Fauville, G., Voşki, A., Mado, M., Bailenson, J. N., & Lantz-Andersson, A. (2024). Underwater virtual reality for marine education and ocean literacy: technological and psychological potentials. *Environmental Education Research*, 1–25.

Flannery, W., Ounanian, K., Toonen, H., van Tatenhove, J., Murtagh, B., Ferguson, L., Delaney, A., Kenter, J., Azzopardi, E., Pita, C., & Mylona, D. (2022). Steering resilience in coastal and marine cultural heritage. *Maritime Studies, 21*(4), 437–446.

Fleming, L. E., Landrigan, P. J., Ashford, O. S., Whitman, E. M., Swift, A., Gerwick, W. H., Heymans, J. J., Hicks, C. C., Morrissey, K., White, M. P., & Alcantara-Creencia, L. (2024). Enhancing human health and wellbeing through sustainably and equitably unlocking a healthy ocean's potential. *Annals of Global Health, 90*(1).

Gates, A. Q., Della-Piana, C. K., & Bernat, A. (1997, November). Affinity groups: A framework for developing workplace skills. In *Proceedings Frontiers in Education 1997 27th Annual Conference. Teaching and Learning in an Era of Change* (Vol. 1, pp. 53–56). IEEE.

Glithero, L.D., Bridge, N., Hart, N., Mann-Lang, J., McPhie, R., Paul, K., Peebler, A., Wiener, C., Yen, C., Kelly, R., & McRuer, J. (2024). Ocean decade vision 2030 white papers-challenge 10: Restoring society's relationship with the ocean.

Gregersen, E. M., Astrupgaard, S. L., Jespersen, M. H., Gårdhus, T. P., & Albris, K. (2023). Digital dependence: Online fatigue and coping strategies during the COVID-19 lockdown. *Media, Culture & Society, 45*(5), 967–984.

Holm, P., Adamson, J., Huang, H., Kirdan, L., Kitch, S., McCalman, I., Ogude, J., Ronan, M., Scott, D., Thompson, K. O., et al. (2015). Humanities for the environment—A manifesto for research and action. *Humanities, 4*(4), 977–992. https://doi.org/10.3390/h4040977

Holm, P., & Brennan, R. (2018). Humanities for the environment 2018 report—Ways to here, ways forward. *Humanities, 7*(1), 3. https://doi.org/10.3390/h7010003

Jefferson, R., McKinley, E., Capstick, S., Fletcher, S., Griffin, H., & Milanese, M. (2015). Understanding audiences: Making public perceptions research matter to marine conservation. *Ocean and Coastal Management, 115*, 61–70.

Jung, J., Gupa, D., Hash, C., Thoms, J., Owens, D., Threlfall, J., & Juniper, S. K. (2022). Doubling down on wicked problems: Ocean artscience collaborations for a sustainable future. *Frontiers in Marine Science, 9*, Article 873990. https://doi.org/10.3389/fmars.2022.873990

Keast, R., Mandell, M. P., Brown, K., & Woolcock, G. (2004). Network structures: Working differently and changing expectations. *Public Administration Review, 64*, 363–371. https://doi.org/10.1111/j.1540-6210.2004.00380.x

Kelly, R., Evans, K., Alexander, K., Bettiol, S., Corney, S., Cullen-Knox, C., Cvitanovic, C., de Salas, K., Emad, G. R., Fullbrook, L., & Garcia, C., (2021). Connecting to the oceans: Supporting ocean literacy and public engagement. *Reviews in fish biology and fisheries*, 1–21.

Mascia, M. B. (2003). The human dimension of coral reef marine protected areas: Recent social science research and its policy implications. *Conservation Biology, 17*(2), 630–632.

McKinley, E., Burdon, D., & Shellock, R. J. (2023). The evolution of ocean literacy: A new framework for the United Nations ocean decade and beyond. *Marine Pollution Bulletin, 186*, Article 114467.

McKinley, E., Kelly, R., Mackay, M., Shellock, R., Cvitanovic, C., & van Putten, I. (2022). Development and expansion in the marine social sciences: Insights from the global community. *iScience, 25*(8), 104735.

McKinley, E., Acott, T. G., & Yates, K. L. (2020). Marine social sciences: Looking towards a sustainable future. *Environmental Science and Policy, 108*, 85–92.

McKinley, E., & Fletcher, S. (2012). Improving marine environmental health through marine citizenship: A call for debate. *Marine Policy, 36*(3), 839–843.

McKinley, E., & Fletcher, S. (2010). Individual responsibility for the oceans? An evaluation of marine citizenship by UK marine practitioners. *Ocean & Coastal Management, 53*(7), 379–384.

Musgrove, M. M. C., Ko, M. E., Schinske, J. N., & Corwin, L. A. (2024). Broadening participation in biology education research: A role for affinity groups in promoting social connectivity, self-efficacy, and belonging. *CBE—Life Sciences Education, 23*(1), ar8.

O'Leary, B. C., Copping, J. P., Mukherjee, N., Dorning, S. L., Stewart, B. D., McKinley, E., Addison, P. F., Williams, C., Carpenter, G., Righton, D., & Yates, K. L. (2021). The nature and extent of evidence on methodologies for monitoring and evaluating marine spatial management measures in the UK and similar coastal waters: A systematic map. *Environmental Evidence, 10*(1), 13.

Ounanian, K., Van Tatenhove, J. P., Hansen, C. J., Delaney, A. E., Bohnstedt, H., Azzopardi, E., Flannery, W., Toonen, H., Kenter, J. O., Ferguson, L., & Kraan, M. (2021). Conceptualizing coastal and maritime cultural heritage through communities of meaning and participation. *Ocean & Coastal Management, 212*, Article 105806.

Overland, I., & Sovacool, B. K. (2020). The misallocation of climate research funding. *Energy Research & Social Science, 62*, Article 101349.

Partelow, S., Schlüter, A., Ban, N. C., Batterbury, S., Bavinck, M., Bennett, N. J., Bleischwitz, R., Blythe, J., Bogusz, T., Breckwoldt, A., & Cinner, J. E. (2023). Five social science intervention areas for ocean sustainability initiatives. *npj Ocean Sustainability, 2*(1), 24.

Popova, E., Aksenov, Y., Amoudry, L. O., Becker, A., Bricheno, L., Brown, J. M., Clare, M. A., Evans, A., Evans, C., Fowell, S. E., & Jevrejeva, S. (2023). Socio-oceanography: An opportunity to integrate marine social and natural sciences. *Frontiers in Marine Science, 10*, 1209356.

Roome, N. (2001). Editorial: Conceptualising and studying the contribution of networks in environmental management and sustainable development. *Business Strategy and the Environment., 10*, 69–76.

Shellock, R. J., Fullbrook, L., McKinley, E., Cvitanovic, C., Kelly, R., & Martin, V. (2024). The nature and use of ocean literacy in achieving sustainable ocean futures: A systematic map. *Ocean & Coastal Management, 257*, Article 107325.

Spalding, A. K., & McKinley, E. (2024). The state of marine social science: Yesterday, today, and into the future. *Annual Review of Marine Science, 17.*

Stojanovic, T. A., Ball, I., Ballinger, R. C., Lymbery, G., & Dodds, W. (2009). The role of research networks for science-policy collaboration in coastal areas. *Marine Policy, 33*(6), 901–911.

Trakadas, A., Firth, A., Gregory, D., Elkin, D., Guerin, U., Henderson, J., Kimura, J., Scott-Ireton, D., Shashoua, Y., Underwood, C., & Viduka, A. (2019). The ocean decade heritage network: Integrating cultural heritage within the UN decade of ocean science 2021–2030. *Journal of Maritime Archaeology, 14*, 153–165.

UNESCO-IOC. (2021). The United Nations Decade of Ocean science for sustainable development (2021–2030) Implementation plan. UNESCO, Paris (IOC Ocean Decade Series, 20.)

van Putten, I., Kelly, R., Cavanagh, R. D., Murphy, E. J., Breckwoldt, A., Brodie, S., Cvitanovic, C., Dickey-Collas, M., Maddison, L., Melbourne Thomas, J., Arrizabalaga, H., Azetsu-Scott, K., Beckley, L. E., Bellerby, R., Constable, A. J., Cowie, G., Evans, K., Glaser, M., Hall, J., Hobday, A. J., Johnston, N. M., Llopiz, J. K., Mueter, F., Muller-Karger, F. E., Weng, K. C., Wolf-Gladrow, D., & Xavier, J. C. (2021). A decade of incorporating social sciences in the integrated marine biosphere research project(IMBeR): Much done, much to do? *Frontiers in Marine Science, 8*, 662350. https://doi.org/10.3389/fmars.2021.662350

Vadrot, A. B., Rodríguez, S. C. R., Brogat, E., Dunshirn, P., Langlet, A., Tessnow-von Wysocki, I., & Wanneau, K. (2022). Towards a reflexive, policy-relevant and engaged ocean science for the UN decade: a social science research agenda. *Earth System Governance, 14*, 100150.

Vince, J., & Hardesty, B. D. (2018). Governance solutions to the tragedy of the commons that marine plastics have become. *Frontiers in Marine Science, 5*, 214.

Weber, C., Golding, S. E., Yarker, J., Teoh, K., Lewis, R., Ratcliffe, E., Munir, F., Wheele, T., & Windlinger, L. (2023). Work fatigue during COVID-19 lockdown teleworking: The role of psychosocial, environmental, and social working conditions. *Frontiers in Psychology, 14*, 1155118.

CHAPTER 3

"Il mare unisce ciò che la terra divide"

Enrico Squarcina

3.1 Paradise Lost?

In the Medici port of Livorno, on a wall stands a blue writing on a white background: "The sea unites what the land divides". The phrase, certainly suggestive, is deeply polysemous (Pic. 3.1).

It may indeed suggest how the sea, which has always been a privileged space of communication, has united distant peoples for trade, knowledge, and unfortunately, domination and exploitation. It may mean that at sea, among members of the same crew, the impossibility of differentiating fates in the event of a shipwreck and the need for cooperation means that, at least partially, social differences are cancelled. Furthermore, it may mean that sharing the experience of the physicality of the sea and the routes that the conformation of landmasses and the general pattern of atmospheric circulation make the same since time immemorial, creates a community that unites sailors of every nation and every time. Likewise, it may mean that frequenting it breaks the rift between the solid surface of our planet and its liquid part by overcoming the terracentric attitude that characterises most human beings. This phrase can be interpreted as

E. Squarcina (✉)
Università of Milan Bicocca, Milan, Italy
e-mail: enrico.squarcina@unimib.it

C. Certomà (ed.), *Blue Kinships*,
https://doi.org/10.1007/978-3-031-78619-8_3

Picture 3.1 "The sea unites what the land divides", Livorno, Italy (Photo Enrico Squarcina)

the possibility that human beings at sea can regain their lost communion with naturalness.

In modern Western societies, humans are becoming increasingly less attentive to the physical and biological elements of the world in which they are embedded. The technology that characterises our way of life has, at least in normal situations, diminished the awareness of being part of the world. Current means of transportation have almost eliminated the sense of distance and the effort required to travel it; the places in our lives are no longer connected to areas or routes with which we engage, but are seen as separate points linked by fleeting temporal gaps filled by scrolling images on screens or any activity that can distract from the experience of the world. Comfortable dwellings and clothing have minimised the impact of climatic conditions and seasonal rhythms. The use of electric lighting can make us overlook the cycle of day and night. The availability of food from all over the world and in every season has led us to view food not as a complex product of natural factors, transformative abilities and cultural elements, but as a standalone entity focused on providing nutrients or gustatory pleasure. All of this results in individuals becoming increasingly unaware of their connection to the natural world, which is now often experienced hedonistically and aesthetically or, when nature shows its power, as an enemy to be destroyed or harnessed. Franco Farinelli (1992) criticised the modern geography attitude, comparing it to a cartographic vision that signifies detachment and non-involvement between the scholar and the world being described, while also highlightening the scholar's active role in shaping the world in his own image. The contemporary attitude towards the natural environment is similar; we deny our connection to nature, seek to control it (or believe we can control it) through technology, interpret it aesthetically, and homogenise it to suit human preferences attributing infinite benevolence or destructive cruelty to it.

3.2 Diversity of the Sea

If this process of physical and cultural domestication is largely successful for the emerged part of our planet, it is much less so for the oceanic part. The sea is conceptually alienating to those who live on land; if one adopts terrestrial concepts and viewpoints, it escapes appropriation.

Lingering on the beach facing the sea, the gaze extends as far as the horizon line, presenting an illusory image of infinity. The idea of the sea's infinity derives, on the one hand, from the curvature of the earth's surface, which in most cases prevents us from seeing its end, whether one is on a beach or on the high seas. On the other hand, as Roberto Casati states, it comes from the confusion between "unlimited" and "infinite". Moving on the surface of a sphere, we never encounter an edge, we find no limits, but this does not mean that the sphere is infinite. Therefore, in the sea, we could sail indefinitely around landmasses, never reaching a shore. Our path may be unlimited but on a surface that is far from infinite (2022, pp. 132–133).

Infinity is by definition limitless; it has no centre, no distance, and no dimensions. In the presence of infinity, or the idea of infinity, the human being loses every possible point of reference. Faced with the infinitely large one feels infinitely small, loses all individuality, and becomes absorbed in the whole. This can evoke feelings of exaltation or anguish that are difficult to manage. According to Roberto Casati the feeling of the sublime emerges, the annihilation of the sense of self in front of an environment perceivedas boundless and non-negotiable (Casati, 2022, p. 65). The infinite, even if only illusory, is not measurable, cannot be graspedby the senses, and since knowledge always starts with the senses, it is unknowable. This leaves room for imagination, meditation or, for those unaccustomed to reading the sea, boredom or rejection. The illusory infinity of the sea fascinates us, interpreted as a metaphor for our desire to transcend human limits, while also unsettling us. The lack of boundaries disorients us.

Angelo Turco (1988) states that human beings need to materially, cognitively, and symbolically appropriate spaces. To do this he enacts the process of territorialisation, which can be schematically divided into three operations: naming, reification, and structuring. The latter relates to the organisational control of space and consists of its delimitation and subdivision. In fact, our author states that humans have a passion and need for divisions to delimit the territorial contexts in which society realises its goals (Turco, 2010, p. 73). Following this line of reasoning, the impossibility of drawing boundaries on the sea, at least boundaries resting on recognisable physical features, prevents one from thinking that contexts can be identified on the sea in which human goals can be realised. Moreover, one of the legal and social foundations of our Western civilisation is private property, applied to objects and the earth's surface. To enjoy the object possessed it must be precisely identified and bounded by borders. Only an object can be precisely identified, and a piece of land fenced off, and thus only a solid part of our planet can full possession be claimed. Claims of sovereignty by states of larger or smaller parts of the sea have always come up against the difficulty of delimiting and controlling it and have had to adapt, allowing exceptions and devising special legislation, to the characteristics of this space which, despite its vastness, constitutes a unitary physical and biological system.[1] The sea therefore prevents the application of one of the principles underlying Western society, that of dividing by having.

When we look at the sea, our eyes observe a changing surface. Its colour varies, even in the space of a few moments; its shape also changes, whether it lies flat as an oily surface or bolts with thunderous fury over the

[1] This feature of the sea is reaffirmed by the first principle of Ocean Literacy, which in fact recites: "The Earth has one great ocean with several features" (https://oceanliteracyitalia.it/sette-principi/).

land, it is mobile. The rhythmic flow of the tides also causes its amplitude to change, covering and uncovering vast amphibious spaces. Invisible to the human eye in the sea move constant or periodic currents, rivers of water in the water, capable of changing climates, aiding, or hindering navigation, eroding, or increasing coastlines. At sea one cannot remain stationary; anchors, dead bodies, and docks can provide some stability only by connecting floating objects on the sea to the underlying earth's crust and thus only near the shore. On the high seas an entirely relative stability is represented by the course, that is, a direction, a becoming, which can constitute if not a point, a line of reference however wavy and sometimes necessarily zigzagging. All this makes one perceive the sea as "different" and indecipherable; stability is a value for human beings. We look for solid ground where we can build ideally imperishable artefacts, we want to interact with people with a character that allows us to predict their reactions, we would desire predictable climatic features, a tamed environment that holds no surprises. Even as far as social life is concerned, stability is favoured; we have idealised settledness, making nomadism practically disappear, towards which the attitude is one of distrust if not open hostility. Simone Regazzoni states that: "Earth is the name of our need for foundation, stability, rational measurability, identity: it is the name of a reality and, more precisely, it is the construction of a reality[2] that tries in every way to protect itself from the fluid, becoming element"[3] (2022, p. 31).

Human beings have difficulty representing, and through figuration making their own, this changing space. The geographical representation par excellence, the map, struggles to represent the sea. Its mutability is generally transformed by the map into uniformity. In most general maps the sea is represented as a homogeneous blue-grey expanse; even when bathymetric hues are used, they refer to the depth of the sea and thus describe the distance between the two limits containing the sea: the atmosphere and the earth's crust, and not the surface or volumetric essence of the sea. The characteristics of the different sections of the sea in nautical charts in which, usually, the sea is depicted as a blank white surface, punctuated by a series of numbers indicating its depth, by acronyms informing about the nature of the seabed, and by isobaric lines joining points at

[2] Italics by the author.

[3] Our translation.

the same depth, not informing, about the changing characteristics of its surface or those of the water mass.

Another element of otherness that we grasp when looking at the sea comes from our perception of it as a near-surface. Being able to penetrate with sight to a limited depth we understand its essence of volume, but of which we can investigate with our senses only a very limited thickness. Rationality tells us that there is something down there; not being able to see it intrigues and disquiets. Only the explorations of the last century have falsified assumptions of motionless, lifeless ocean floors. The vacuum of knowledge has allowed the imagination to populate the deep sea with fantastic and monstrous beings of all kinds. Perhaps the continuation for so long of ignorance about the ocean depths and the impossibility of probing its depths with the senses, and for so long with instruments, has left its mark on the way we think about the sea.

The sea is a vast volume containing life that is alien and largely unknown to us. Roberto Casati (2022) asserts that the sea is both a container and content whose form, like that of any liquid, is determined by its container (49). Moreover, he asserts that we do not even know what a definition of the sea should refer to—the mass of water alone or the biomass it contains, in addition to everything that is not water. According to this author, the sea is not a living thing, but is everywhere full of life (57–58). It is not a desert, but a mass in which organic and inorganic elements mingle, further undermining the need for discrete subdivisions.

At sea, offshore, there are no fixed structuring elements of the landscape, so reference points on its surface are lacking. Those on the high seas feel like they are at the central point of a circle that extends the same in all directions and does not change as the boat moves forward. The perception of motion is relative; we realise we are heading in one direction more or less quickly as we see the position of objects around us change. When sailing on the high seas, where there are no fixed objects to show our motion, if our gaze does not rest on the prow plying the sea or on the wake left by the boat, we have the feeling that we are not moving. We feel enmeshed in a liquid world from which there is no escape, imprisoned in an anguished timeless space.

Until the adoption of the compass and later the sextant and chronometer, and more recently, radio or satellite geolocation systems, orientation was mainly done through the observation of features outside our planet or through sensory perceptions. For example, sailors would look for towering clouds rising in the sky at distant islands and lands, or they would perceive the smell of the earth as they approached it. They would also observe the flight of birds that did not venture into the open sea. Experience taught them that depending on latitudes, seasons, and times of day, winds predominantly blew from a certain side, and consequently, the background wave came from a more or less constant direction. Conversely, the presence of an island or shallow water would disrupt the regular succession, height and period of waves. These empirical and sometimes deceptive modes of orientation required constant attention to the environment in which the sailor was navigating. Matteo Aria reports, according to Jake Phelan, that the ability to sail, especially under sail, "would demand the engagement of all the senses to relate to the stimuli provoked by wind, sky and sea, making the experience of sailing something closer to sensory and practical involvement, rather than the enactment of mechanically embedded cognitive baggage. Going to sea would thus mean attuning one's senses to the environment, harmonising perceptions and learning to atmospheric changes, variations in gusts, and the rhythm of the waves"[4] (2021, p. 18). This immersion in nature is becoming increasingly rare, in high-tech countries.

3.3 Limits and Freedom

Sailing by exploiting wind propulsion corresponds to getting directly in touch with nature. The sea unites with our primordial element that the modern terrestrial condition has obliterated. The pretence that we can overcome nature, that we can nullify it by will and technology, is soon annihilated in the presence of the sea. Navigation is an activity of continuous mediation between the destination, the weather conditions, the state of the sea, and the need to keep within safe limits. It is not a matter of overcoming nature, but of adapting and acting accordingly to what it proposes. The sea may dictate that we choose a longer course to seek the little wind it brings forward, it may dictate that we postpone a departure or seek shelter or even reverse course to avoid a collision with its unleashed power. Nature imposes choices; to be able to choose is to be free. Certainly, it is not about absolute freedom, a utopian conception of romantic and/or superhuman derivation, an illusion created by the estrangement from nature, by having made technology

[4] Our translation.

the modern deity. Éric Tabarly, one of the most representative navigators of the recent past, wrote that the boat, contrary to what some people think, is not freedom, since sailing means accepting the constraints that the sea imposes; however, these are constraints resulting from a voluntary choice, and this is a privilege, since most humans suffer obligations that they have not chosen (Tabarly, 1997, p. 98). He thus states that freedom consists in choosing the rules to which one submits.

Freedom is connected to free will, and this implies that only those who freely choose to set out to sea, certainly not those who are forced to embark to reach a shore that promises a better life or because no other job prospect is offered to them, can accept its rules, can find there the sense of freedom, can find there that physical and mental connection with the natural part of ourselves, often latent but always ready to emerge.

The identification of the sea with freedom can also emerge indirectly, without personal experience of the sea. Jean Le Cam, a veteran of ocean sailing competitions, reached by the news that he had incurred a penalty for crossing a vessel traffic exclusion zone at a regatta, said in direct satellite link, "If there is no possibility of doing it, then put mandatory points on it. In any case, there are only mandatory things in this low world [...] Why are we here? Because we make people dream. Because they have too many prohibitions. Obligations are part of everyone's daily life. We must stop with the prohibitions. This Vendée Globe and the sea are still our only spaces of freedom. Please don't spoil it. Thank you!".[5] This outburst introduces well the theme of projection by the public, perhaps even among those who have never sailed, of the idea of freedom on the sea. I think it is impossible to take a census of narratives in which the sea is implicitly or explicitly related to freedom. As we have seen, there is a substantial difference between the sea environment and the land environment; this can instil fear or indifference. It suggests at the same time the idea of a space in which the rules of life that take place on the solid surface of the planet do not apply, in which it is possible to escape from social obligations, in which the dream of absolute freedom can be achieved.

However, for most people, the sea remains a mythical space filled with incomprehensible and uncontrollable primordial forces. Those who frequent it are seen as special beings, heroes who are admired but cannot be imitated. Literary and cinematic narratives set at sea often reach their

[5] Our translation from http://www.dailymotion.com/video/xvalmg_jean-le-cam-nous-parle-des-penalites_sport#.Uc1YtYNH7cs (last viewed 01/03/2024).

climax during storms (Farnetti, 1996, p. 175), depicted as unpredictable events that disrupt the plot and to test the protagonists. According to Michel Roux (1997) these episodes are not just descriptions of weather events but are meant to reveal the characters of the protagonists, showcasing their heroism or cowardice. The fascination with modern ocean races lies in the challenge they present to the mythical space of the sea, particularly its dangers like storms. Many times, the heroes in these races end up as losers, struggling against the gods or nature itself (which from positivism onward has been elevated to the status of deities). This doomed struggle often leads them to a superhuman condition. Even modern sailors who disappear at sea can be considered contemporary heroes, demonstrating of the need for extraordinary abilities to face the mythical space of the sea. In the walled city of Saint Malo, the starting point of the most famous and followed trans-ocean solo races, the Route du Rhum,[6] a plaque commemorates the tragic end of: "Alain Colas, navigator and honoured citizen of the city, left the city on 'Manureva' on the 1978 Route du Rhum, it was his last voyage",[7] constituting a warning to those who want to face the ocean and at the same time placing the unfortunate sailor in the role of the heroes of the sea. Éric Tabarly's own disappearance at sea, which occurred in the Irish Sea on the night of June 12–13, 1998, helped to increase his aura of heroism and build the idea of heroism attributed to sport sailors. If before the fatal accident he was considered one of the greatest racers, a sailing genius, admired for his victories and innovative skills, with his death he took on the tragic character of ancient heroes, someone inimitable, with unmatched qualities. The interest in deep-sea sailing and its protagonists, along with ancient and modern narratives, paradoxically creates a distance between humans and the sea. It fails to unite people with the sea and does not achieve the goal of bringing humans and nature closer together.

[6] Route du Rhum is a solo, trans-oceanic sailing regatta without assistance that takes place every four years and connects Saint Malo to Pointe à Pitre in Guadeloupe.

[7] Our translation.

3.4 Aesthetisation and Indifference

The consequence of this estrangement is, on the one hand, the aestheticisation of the sea, which persists except as the background of the now-dying picture postcards, in the backgrounds of selfies (complete with palm trees and setting sun) and pictures in tourist brochures, and on the other hand, indifference to the sea beyond the coast.

Indifference that has allowed, despite the fact that the planet's liveability depends largely on the presence of the sea, to make this space the dumping ground for the leftovers of a consumerist civilisation, to consider it an inexhaustible source of mineral and biological resources that are being depleted, a void to be overcome in order to connect places on the Earth's solid surface.

We must therefore rebuild the connection between humans and the sea that the writing on the wall in Livorno claims and, more broadly, between humans and nature.

If human geography has as its purpose the study of the relationship between humans and their host planet, it should play an important role in bringing out and fostering this connection.

Concerns over the global environmental crisis have brought to attention the need to preserve the sea for the preservation of the planetary environmental balance, also demonstrated by the United Nations General Assembly's proclamation, for the period from 2021 to 2030, of the "Decade of Ocean Sciences for Sustainable Development", in this wake, debates and scientific research on the marine world are multiplying, actions for mitigation and adaptation to the rising average sea level are being sought, it seems, however, that all of this is part of a separation and subjugation of the sea to the needs of the earth. We seem to be able to read between the lines that the idea of ocean sustainability consists of preserving ocean spaces to allow life on land, or worse, to perpetuate the current, consumerist, rapacious, and disegalitarian/elitist model of development on the emerged part of the planet. After all, it is okay, that something needs to be changed for everything to stay the same.

Instead, it is necessary to change the way we think about the sea and our relationship with it, to understand the need for unity that the Labronian wall once again emphasises. Education can play an important role in this direction. While knowing the biological and physical aspects of the sea can help us understand it, it may not necessarily change attitudes towards it. Therefore, an emotional as well as cognitive connection is needed. We need to make people feel that the sea is a part of us, to overcome the terracentric cultural construct. As Simone Regazzoni declares, "a paradigm shift that questions our fundamental beliefs that deconstructs the earth as the 'quintessence of the human condition'"[8] (2022, p. 36). Terracentrism closely linked to anthropocentrism, assigns value to nonhuman entities only based on human interests and emotions. It turns the beach into a privileged playground and the vast expanses of the high seas into ignored or misrepresented spaces. It attributes feelings of love and friendship to colourful tropical fish and sea predators, as well as to storms and malevolent forces, and distancing us from the possibility of envisioning equal coexistence on the planet with all existing forms of life (Latour, 2020).

If territory is the product of a process of appropriation of a space to comply with human needs, it is necessary to realise that these needs are not only about material needs. Equally important is to create an affective bond with the spaces one lives in. To live well, we need to conceive of a different relationship with our planet, one that does not think of the natural elements, and according to our reasoning, the sea, as amorphous, detached matter with no value other than that of utilitarian space, but starts from an awareness of the need for an alliance with it, of the need for a balanced and equal relationship. Affective territorialisation is needed, changing the conceptualisation of the sea from space to place.

Human action on and in the world derives not only from actions dictated by rationality, but also from the values we attribute to its parts. Adalberto Vallega asserts that cultural geography "can be delineated as a field of study that focuses its attention not on territorial reality as such, but on the symbols that human communities have attributed to places and the meanings and narratives that are connected to these symbols"[9] (Vallega, 2002). The study of the values attributed to marine spaces by the human community can therefore be of geographical interest and fit fully into the field of cultural geography. The study of the values, names, and meanings that different cultures have over time attributed and attribute to the sea, would not respond to a simple intellectual curiosity, it would provide one of the keys to understanding human action on the sea. It could make us aware that another vision of the relationship between humanity and

[8] Our translation.

[9] Our translation.

ocean spaces is possible. The appropriation of the world, the transformation of it from amorphous matter to territory, involves, as Angelo Turco (2010) teaches us, not only its physical transformation and structural organisation, but also a symbolic, that is, cultural appropriation. A cultural geography of the sea, understood as the study of symbolic relations with this space could help understand the deep needs of human beings including the need for rapprochement, not hierarchical but equal, with the complexity of the planet in which they live.

The challenge is complex; it is not only about changing the way we think about our way of being on Earth, but also about taking into account the results of the interactions between the human and natural spheres. In fact, since the early 2000s human beings have been recognised as an active part of Earth's evolution, so much so that our time has been identified in a new geological era, the Anthropocene. However, there still remains an idea of separateness between humanity and nature, especially the belief that the active role in the transformation of the world belongs solely to humans, for better or worse. The result is the inability to relate to the new life forms born from our actions on Earth and the impossibility of imagining coexistence with them.

Donna Haraway (2019) warns us that the changes, of the chemical, physical, and geological characteristics of the Planet brought about, voluntarily or involuntarily, by human action and the input of waste products of our civilisation, have left irreversible marks to which our nonhuman cohabitants have reacted with new evolutionary forms, becoming active factors in the transformation of the planet. This is a reality with which our way of conceiving the world must come to terms, not only from a physical point of view but also from a conceptual one, and which leading to a new conception of the human relationship with the nonhuman, the other-than-human, of which the sea is perhaps the most obvious epiphany.

The phrase quoted at the beginning of this contribution, "The sea unites what the land divides", can and should therefore be interpreted not only as a statement but as a wish, as a program of understanding and coexistence with the other and the elsewhere.

References

Aria, M. (Eds.). (2021). Ermenautica. Dai mari condivisi i segreti della convivenza, Novate Milanese (MI): Prospero Editore.

Casati, R. (2022). *Oceano*. Einaudi.

Crutzen, P. (2002). *Benvenuti nell'Antropocene. L'uomo ha cambiato il clima, la Terra entra in una nuova era*. Mondadori.

Farinelli, F. (1992). *I segni del mondo. Immagine cartografica e discorso geografico in età moderna*. La Nuova Italia.

Farneti, M. (1996). *Il romanzo del mare. Morfologia e storia della narrativa marinara*. Le Lettere.

Haraway, D. (2019). *Chthulucene. Sopravvivere su un pianeta infetto*. Nero.

Latour, B. (2020). *Politiche della natura: Per una democrazia delle scienze*. Milano, Raffaello Cortina.

Regazzoni, S. (2022). *Oceano. Filosofia del pianeta*. Ponte alle Grazie.

Roux, M. (1997). *L'imaginaire marin des Français: mythe et géographie de la mer*. l'Harmattan.

Tabarly, É. (1997). *Mémoires du large*. Éditions de Fallois.

Turco, A. (1988). *Verso una teoria geografica della complessità*. Unicopli.

Turco, A. (2010). *Configurazioni della territorialità*. Franco Angeli.

Vallega, A. (2002). Luoghi, spazi, simboli, Università di Catania: Facoltà di lingue e letterature straniere, 20 XI 2002, www.vallega.it

CHAPTER 4

Thinking-with the Sea's Turbulent Materiality: Trans-Oceanic Figurations, Blue Kinships and Liquid Futures in Lagoon by Nnedi Okorafor

Gabriella Palermo

4.1 Critical Ocean Thinking: Introduction

In recent decades, the Oceanic Turn (DeLoughrey, 2016;Steinberg, 2001) and the Critical Ocean Geographies (Anderson & Peters, 2014; Peters et al., 2022; Steinberg & Peters, 2015) have repositioned the sea as a space at the centre of social and humanistic disciplines. Rather than a blank space or a platform of mere crossing, ahistorical and atemporal, as it has long been perceived especially in the Western thought, the oceanic perspective attempts to think-with the sea as a tool for plural, fluid, continuously changing becoming-with that helps us rethink and develop other relations, other politics, other possibilities. In particular, Peters and Steinberg's (2019) more-than-wet geographies, think-with the

G. Palermo (✉)
University of Palermo, Palermo, Italy
e-mail: gabriella.palermo@unipa.it

C. Certomà (ed.), *Blue Kinships*,
https://doi.org/10.1007/978-3-031-78619-8_4

sea starting from its ontology, that is, from its different materiality, made of extension and excess: the sea extends beyond the liquid spaces of a single sea-space, in a trans-oceanic (Palermo, 2024) and trans-actional connection (Palermo & Steinberg forthcoming); the sea is an excess of its materiality—as it exists in the liquid, solid and gaseous state—and is present in the constitution of a pluriverse of relations, thus facilitating imaginative transformations and the construction of other representations (Peters & Steinberg, 2019).

In connection and alliance with the new feminist and posthuman materialism, a paradigmatic shift elaborated especially by figures such as Haraway (1991), Braidotti (2013), Barad (2007), Alaimo (2010), Neimanis (2019), among others, observing matter and the incessant materialising of the world helps us to revise 'notions of change, agency causality, time space for ecologically ethical positionings that unite humanities, literature, natural sciences, research practices, representations, aimed at politically intertwining nature and life from postcapitalist, posthuman, postcolonial perspectives' (Borghi, 2017, p. 7, my trans.). In this materialist onto-epistemology, multiple knowledges converge to elaborate approaches and relational modalities in which matter and metaphor, fact and fiction, materiality and discourse decentralise the gaze of the unique human subject—which is, the human male, white, Western, able-bodied, heterosexual power subject of capitalism, as inseparable from extractive colonialism.

What meanings does thinking about the living and ever-changing matter in death, about the agency and power of the materiality of the existing of everything around us, bring with it?

> Why advocate the vitality of matter? Because my hunch is that the image of dead or thoroughly instrumentalised matter feeds human hubris and our earth-destroying fantasies of conquest and consumption. It does so by preventing us from detecting (seeing, hearing, smelling, tasting, feeling) a fuller range of the non-human powers circulating around and within human bodies. [...] The figure of an intrinsically inanimate matter may be one of the impediments to the emergence of more ecological and more materially sustainable modes of production and consumption. My claims here are motivated by a self-interested or conative concern for *human* survival and happiness: I want to promote greener forms of human culture and more attentive encounters between people-materialities and thing-materialities. (Bennett, 2010, pp. IX–X)

But what if these more-than-human encounters Bennett speaks of as vibrant matter, instead of green, were blue? What alternative politics could be developed if we change our perspective and move away from Terracentrism to turn our gaze to the sea? What policies, relationships, possibilities, ecologies, kinships, does the sea suggest us through its materiality? And how do blue ontologies and epistemologies help us think, imagine and practise alternative worlds and futures against extractive and colonial capitalism?

In this chapter I[1] try to put the conceptualisation of the Oceanic Turbulent Materiality (Steinberg & Peters, 2015; Palermo, 2024) to work with the Blue Humanities and posthuman feminist New Materialism, as discussed in the second section. In the third section, these are interwoven with Nnedi Okorafor's novel *Lagoon* (2014), in which the alternative to capitalist extractivism is positioned in the west coast of Nigeria, the neuralgic reservoir for rethinking worlds outside the colonial violence of racial capitalism.

Why a science fiction novel to materialise the entanglements and possibilities of the Blue Humanities allied with Critical Ocean Geographies?

In the 1999 interview *How like a Leaf*, Donna Haraway describes bioscience as a discipline built in the tensions between matter and metaphor: we live both 'as' and 'in' a world in which biology is part of a discursive and material knot of matter and metaphor. This discursive and material knot is exemplified by Karen Barad in the figure of the ophiuroid, a sea creature widespread in the Mediterranean area belonging to the class of Echinoderms. Similar to the starfish, this sea creature expresses a morphology in which being and knowing, materiality and intelligibility, substance and form are intimately entangled and intertwined (Lehman et al., 2021). Through this sprawling being, with the capacity to regenerate itself, Barad speaks of the relationship between matter and meaning, which takes shape through the relationship with a world that, like the ophiuroid, is in constant becoming (Barad, 2014). This interweaving of matter and representation is exemplified by Donna Haraway in her book *Chthulucene* (2016) through FS figurations, where the acronym stands for Scientific Fact, Science Fiction, Speculative Fabula, String Figure, Speculative Feminism: together, they represent figurations for imagining and living in the Chthulucene, the world that was, that could be and that will

[1] An initial core of reasoning further elaborated in this chapter can be found in Palermo (2024).

be, in the constitution of multi-species kinship beyond the violence of extractive capitalism. In the ecological crisis we are living in, instead of developing only vocabularies of catastrophe, we need to introduce invention in our existence: situated more-than-human speculation are already collaborative practices to practice possible collaborative futures.

The chapter concludes by weaving these knots together and suggesting how the Blue Humanities, in alliance with Critical Ocean Geographies, represent paradigms and tools not only for repositioning the sea at the centre of our geographies in order to reconfigure them, but also for attempting to develop alternative and responsible knowledge and politics.

4.2 A Turbulent Materiality

Starting point and central paradigm of the Oceanic Turn is that the sea has a different materiality: fluid, liquid, extensive and excessive, never static, constantly becoming, constantly co-digesting and co-composing forces and materials, this more-than-wet materiality questions us, disturbs us, invites us to engage with it in this turbulence.

In *Turbulent Waters in Three Parts* (2021), Jessica Lehman, Elizabeth Johnson and Philip Steinberg, argue that there is no concept that more accurately encapsulates the Oceanic Turn than turbulence. They report, that according to the *International Science Council's Scientific Committee on Oceanic Research*, turbulence is a condition of fluid flow in which each component of velocity and vorticity is irregularly distributed in both time and space: because of these characteristics of change, movement and chaos, it is a cherished feature of the ocean geographers.

How does the ocean introduce turbulence into historical narratives? How is turbulence both disruptive of and encoded in the legal territorialisation of the ocean? What are the linkages between turbulent knowledge and turbulent nature? What does turbulence have to do with the relationship between living on and with the sea? At what scales does turbulence matter? And, ultimately, how can a turbulent approach to oceanic knowledge teach us about ourselves as (partially) marine creatures? (Lehman et al., 2021, p. 4).

Lehman et al., (2021) definition of turbulence as a typical characteristic of the space of the sea, refers in my view to meanings discernible in the Harawayan 'Trouble', as a space for experimentation with alternative models and relations from which fluid, moving, three-dimensional geographies and politics can emerge, as opposed to univocal, flat, binary and dichotomous representations. As a matter of fact, in *Staying with the Trouble. Making Kin in the Chthulucene* (2016), Donna Haraway starts from the Trouble to point to the critical state of the planet that necessitates the articulation of responses (the respons-ability), to be built through multi-species narratives and sympoietic models of being-with, for the imagination of new worlds and the construction of alternative possible futures. As the introduction to the text emphasises, Trouble comes from the thirteenth-century French verb *Troubler and* means 'to stir', 'to make opaque', 'to disturb'. The Italian translation of the text in 2019 (*Chthulucene. Sopravvivere su un pianeta infetto*) has raised quite a few criticisms, especially for the choice of the translation of 'Trouble' with 'Problem', a word that seems to obscure this complexity of the knot between matter and meaning. Indeed, it seems that Haraway was mostly emphasising the terms of a 'turbulence' of the planet with which to remain in contact—a term probably closer to disturb, to stir—in order to live and die well in a sympoietic model in the *Chthulucene*—the era that was, that is, that could be, a futuristic time. As Pinto (2021) reports, trouble is a concept that speaks of 'opacity' and refers to several genealogies. On the one hand, to Judith Butler's deconstruction of gender binarism in *Gender Trouble* (Butler, 1990). On the other, as Timeto (2020) argues, it comes from the Latin word *turba*, from the verb *turbulare* and the noun *turbula:* a small disorderly crowd, an entanglement in sympoietic processes.[2] A turbulence, in my opinion, suggested by the sea and its excessive and extensive constituent more-than-wet elements: volumes, liquidites, circular temporalities, tides, waves, wake, sedimentations. They exist in a pluriverse of states, of movements, of more-than-human relations, breaking down dichotomies and decentralising the gaze to a multiple world, in continuous becoming-with; they mix, disturb, amplify other stories through regener-a(c)tions and intra-actions.

'Intra-actions' is a term borrowed from Barad who describes them as 'matter's lively and co-generative agency and lends itself well to ocean relations, [an] intra-acting from within and as part of the world in its becoming' (Barad, 2007, p. 396). As Johnson reports, unlike the interaction between individuals and pre-existing subjects, intra-action refers to those processes in which subjects, objects, forces, elements, collectively materialise as *things-in-phenomena*. And things-in-phenomena for

[2] See also the afterword by Balzano et al. (2022) to the Italian translation of *Making Kin. Making Kin, Not Populations* by Clarke and Haraway.

Barad are what is conditioned by a type of turbulence she calls diffraction. In quantum physics, diffraction denotes the curvature of light waves around an object, referring not only to the wave properties of light, but also to the act of observing light as a wave and not as a particle: unlike reflection, which is concerned with the similarity between things, diffraction requires the recognition of the intertwining of matter, observer and apparatus (Lehman et al., 2021, pp. 23–24). Diffraction is a method for thinking with the turbulent materiality of the sea.

And what meanings does the assumption of this onto-epistemological perspective bring with it?

First, the turbulent materiality of the sea challenges the legibility of space as a neutral medium to instead help us think of it as Massey (2005) put it, i.e. as a space of *stories-so-far*: a plural space, in which multiple spaces, subjects, inhabitants, points of view, stories are woven together, breaking down a world of powers of dichotomies, binaries, vertical powers. In this perspective, the sea's materiality shapes a space co-composed by diverse forces. On the one hand, as a pivotal space for capital, that has always used it since the birth of long-distance trade in the triangular route of Modernity, for the reproduction of violence in the wake (Sharpe, 2016) of racial and extractive capitalism. On the other hand, as every space in which capital accumulates for the sake of value producing death and 'valuable' lives, the space of the sea is also a space for resistance and reproduction of stories and relations that have and could challenge the mode of production, its violences, its hegemonic narrative. Indeed, the value of the ocean as a theoretical and practical 'machine' lies not 'in its existence as an object of alterity (whether real or imagined) but in the ways in which its materiality intersects with global political economies and territories, constructing a "world interior of capital" that both facilitates and disrupts the flows that constitute expansive capitalism' (Peters & Steinberg, 2015, p. 19). Mobility and migratory conflicts, political ecologies and economic socialisations from below, the emergence of Black Undrowned (Gumbs, 2021) bodies and histories in the abysses from the Atlantic to the Mediterranean, the violences of deep-sea mining, are some of the multiple clashing forces, bodies, communities, relations, inhabitants, dominant and counter-archive that co-compose the turbulent materiality of the sea always in motion. Thus, as a multiple space co-produced by *stories-so-far*, not only the sea is a space, but it also suggests a diverse conceptualisation of spatialisation where space is no more a Euclidean geometry of lines, points, areas (Steinberg & Peters,

2015), but is instead made-with flows, voluminous spheres, movements, intra-actions.

On a second level, the turbulent materiality challenges the capitalist conception of time, a progressive straight line on which bodies and territories are consumed as the colonial Other, not in line, not in step, in delay with progress. But the sea has a diverse temporality, which is circular, where everything is continuously in motion, always present, always past, always future. It is the temporality of the tides that Brathwaite (1975) called *tidalectics*, in which the past is always there, in the presence of memories and molecules, shaping the present and co-compounding the future, in an incessant regeneration. Sediments, fragments, detritus, human and non-human bodies, lives and deaths, histories and memories become-with in this fluid archive that always returns everything back. As wrote by Derek Walcott, the sea is history; but the sea is history and geography (Steinberg, 2022) as an archive that is continuously molecular shaped in the sea-space through processes of co-digestion and transcorporeal becoming. Positioning herself in the thought of feminists such as Rachel Carson (1951), who already insisted on the possibility of a transcorporeal connection between the sea and the human being (through the presence of the oxygen of the plankton in our blood and in each of our breaths) Alaimo delineates trans-corporeality as a chain, which needs collective breaths:

> A transcorporeal, oceanic ecocriticism floats in a productive state of suspension, between terrestrial human habitats and distant benthic and pelagic realms, between the aesthetic estrangement of sea creatures and the recognition of evolutionary kinship, between mediated, situated and emergent knowledges and an ethico-aesthetic stance of wonder. Both new materialisms and blue-green environmentalisms suggest that there is no solid ground, no foundation, no safe place to stand. Like our hermaphroditic, aquatic evolutionary ancestor, we dwell within and as part of a dynamic, intra-active, watery world (Alaimo, 2012, p. 490).

In *Bodies of Water* (2019), Neimanis similarly constantly calls for the need to create new imaginaries in relation to water from the principle of incorporation: recognising that we are bodies of water suggests processes of relation that challenge individualism; in which each is thought of as a single entity sufficient into itself; phallocentrism, the empire of non-relation except for power relations; and anthropocentrism, the conceptualisation of the superiority of man against nature. In this hydro-feminist perspective, the body of water exceeds and extends itself in a more-than-wet, relational, imaginative spatiality and materiality. Central to this *matterphorical* (Lehman et al., 2021) oceanic thought are figurations, those transformative cartographies that Braidotti defined 'as "living maps" that acknowledge concretely situated historical positions. Figurations are keys for imagining and living otherwise, but unlike a concept unfettered by the world we actually live in or as, figurations are importantly grounded in our material reality' (Neimanis, 2019, p. 5).

These figurations can represent a point of fall and meeting between the new geographies of the *more-than-wet* ontology and the *Material Turn of* feminist knowledge and practices for the imagination and construction of other possible alternative futures. In fact, they depict a liquid world, constantly changing, plural and sympoietic, and therefore turbulent. Fact and fiction are thus amalgamated in the dominant archives, as embodied by Barad's ophiuroid, where *fiction* becomes fact, and in the counter-archives of the sea, it is *fiction* that incorporates itself with matter and materiality. Through imagination, stories compose alternative worlds for accountable processes. This is why we need stories: to imagine how the world might be otherwise; to decentralise our gaze and broaden our viewpoints; to weave a multi-species collective plot that, on the one hand, unravels the knots of violence of the hegemonic single story of racial capitalism; on the other, makes new ones in which collaborative and accountable worlds are thought and thus practised—in which, that is, we enable each other to generate responses, in a response-able way. In other words, this is way.

> We need narratives of a 'new' kind; narratives to restore the earth and the commons that capitalism has deprived us of. Tales that invite us to regain and create trans-species sensibilities, trans-material vitalities and trans-cerebral motions. […] Because fabulation is a truly new genre of construction, at least for those and those who create knowledge. According to us (who are inhabited by sociological, anthropological and artistic practices), fabulations are those tales that excavate interstices in our world, to provoke new connections and force us to reopen our search, to explore again this abandoned territory that did not seem to deserve our attention.

> It is an act of repopulation that will no longer be trapped by the question of True and False.[3]

These specific stories—as 'it matters what matters we use to think other matters with; it matters what stories we tell to tell other stories with; it matters what knots knot, what thoughts think, what descriptions describe, what ties tie. It matters what stories make worlds, what worlds make stories (Haraway, 2016, p.12)—are partial and situated, and must therefore mobilise oppressions and privileges, in recognition of the different hierarchising positions we occupy in the world, where we are not all responsible in the same way, as Anthropocentrism seems to suggest, nor do we all suffer the violence of capital in the same way, by species, class, gender, race'.

By reconfiguring our conceptions of space and time, hence of history and stories, of bodies and possible relationships, of lives and deaths, of different responsibilities and postures, oceanic turbulent materiality becomes matter and representation, practice and imagination, discourse and politics, science and narration, inviting us to engage with its liquidity.

4.3 Lagoon

Set in Nigeria, in Lagos to be exact, Nnedi Okorafor's *Lagoon* (2014) immediately positions itself in that Africa imaginative reservoir and generator of worlds, as *Africanfuturism* would have it. If the title of the book refers to the colonialism of the past, when in 1472 the Portuguese arrived in Nigeria, the naval 'vanguard' of European expeditions (Lagos in Portuguese means in fact lagoon), the plot of the novel is built around the coloniality of contemporary extractive capitalism, referring in particular to the extraction of oil in the Niger delta by Western hands, with all the production of death that this entails, human and non-human.

How to imagine a transition out of this mode of production? Nnedi Okorafor's imaginative answer is the arrival of shape-shifting aliens in the ocean waters, with the ability to reorder matter (Jue, 2017).

The genesis of the novel can be traced back to the writer's viewing of *District 9*, a 2009 South African film directed by Neill Blomkamp. In this film, the monsters who demonstrate the return of a past that has never

[3] Haraway D, call for papers of the conference *Sympoièse, SF, embrouilles multispécifiques*, Nice, 2013 (my trans.).

passed are shrimp-aliens who arrive in Johannesburg townships, where the 'inclusion' of the monsters takes place through illegalisation:

They, like the ex-immigrants or their children in the Parisian banlieues of segregation and permanent police action, contribute to the state's economy of containment and exploitation of the district's illegal trade and labour force, and to the legitimate use of force for the eviction and gentrification of the township area. On the other hand, the shrimpers [...] do not stay: they rebel by claiming their agency, their right to stay, or to leave (Giuliani, 2015, pp. 73–74, my trans.).

Okorafor criticises the film's portrayal and stereotyping of Nigerians as unscrupulous arms and drug traffickers and chooses to imagine *Lagoon*, the plot in which aliens arrive in the waters of the Lagos lagoon not to colonise it, but to regenerate the world (Colliva, 2021; Vallorani, 2017).

Dedicated 'to the diverse and dynamic people of Lagos, Nigeria—animals, plant and spirit', the novel's opening is firmly set in the belly of the Atlantic Ocean, which we observe through the perspective of a female swordfish seeking revenge: she has decided to fight the effects of extractive capitalism and take her revenge and is headed for a pipeline in the ocean to destroy it.

Immersed in its perspective, we know that the 'new people', the aliens from the ocean, have arrived, who, convening a kind of trans-oceanic assembly of the future, are fulfilling the wishes of the inhabitants of the sea. An octopus gets a damaged tentacle fixed, while everything around it begins to change. Even the swordfish is then persuaded and, as it changes shape and colour, makes its wish, asking to become three times bigger and twice as heavy: 'Now she is no longer a swordfish. She is a monster' (Okorafor, 2014, p. 6). At the end of the assembly with its inhabitants, the aliens have regenerated the ocean as a place of new ecologies, of monstrous desires and alliances: 'Despite the FPSO Mystra's loading hose leaking crude oil, the ocean water just outside Lagos, Nigeria, is now so clean that a cup of its salty-sweet goodness will heal the worst human illnesses and cause a hundred more illnesses not yet known to humankind. It is more alive than it has been in centuries, and it is teeming with aliens and monsters' (*ibid.*).

The novel is divided into three sections: *Welcome*, *Awakening* and *Symbiosis*. In the first section, on the mainland, we find at the Beach Bar, as midnight approaches, Adaora, a marine biologist whose marriage is falling apart; Chris, who has become part of a group of religious fanatics; Anthony, a well-known Ghanaian rapper, who has just finished one of his concerts; Agu, a soldier of firm principles and values, who has just beaten up a superior officer who was raping a woman. At the bar, while each one is weaving his thoughts, a wave coming from the Atlantic Ocean overtakes them, engulfs them and shows and whispers the oceanic world to them. They thus meet what they call Ayodele, an oceanic creature who shows herself to them in the body of a woman, but who is actually capable of changing form and mutating matter: she is a liquid being, co-composed by the fluidity of the sea. Ayodele comes to them in this form because she is looking for human companions who can help her spread her people's message: the promise of a future of a regenerated Lagos, starting with the sea, which must be cleansed of the death toll that oil extraction is causing.

Awakening is instead the part where we observe the city's reactions to the message of the arrival of the shape-shifting aliens, of violent reaction and chaos on the one hand; of hope and desire for change on the other. The fanatical religious groups led by Father Oke—who accuses both Ayodele and Adaora of being sea witches—and the army and politicians, including the President, are portrayed as the conservative 'stop the history' groups, but also embodiments of blind, violent, obtuse and corrupt power that thinks only of its preservation. On the other hand, our protagonists and a Lagos anxious about Nigeria's future and on the move for social change, like that of the LGBTQIA + community, mobilise to create other ecologies and new systems of coexistence, as suggested by the oceanic aliens.

In the *Symbiosis* part, humans, animals and ocean creatures ally themselves to build a world outside extractive capitalism, free from the destruction caused by oil dependence, whose key is the construction of a trans-oceanic environment: a world of symbiosis, i.e. of sympoietic cohabitation, of multi-species kinship, whose key is the regeneration of the ocean. The possibility of alternative futures is thus revealed in the materiality of the sea: an *Alien Ocean* (Helmreich, 2009) that bestows liquid and marine powers on our protagonists, becoming a regenerative space.

In the novel, marine biology and *Africanfuturism*, facts and fabulations, meet to construct an alternative ecology, in a relationship between matter, materiality, resources and human and non-human actors, where the question of oil extraction operates in the geographies of the submerged and the underground of the hydro-social marine world (Phillips, 2018; Zurita et al., 2018). The encounter between humans, aliens, animals and deities does not create a fracture or a separation of foreignness between an inside and an outside, but a relationality, a great hank game, indeed, sprawling filaments and weavings, which Jue (2017) calls *intimate objectivity*, a feminist practice and epistemology embedded especially by Adaora, the marine biologist who weaves fact, fiction and folklore to construct the new ecology of the future city. Among diverse figures of Nigerian cosmologies present in the plot of the novel, Udide Okwank, an Igbo divinity, in *Lagoon* is the spider-weaver of the story (see Serrano, 2020; Colliva, 2021). Always present but never involved in the plot, Udide adopts from time to time in the narration in the third person singular the point of view of Adaora, Agu, Anthony, the female swordfish, the bat or the spider. We only realise in Chapter 44 that Udide is the originator of the narration, when it speaks in the first-person singular (and in italics):

The sea always takes more than it gives. Right now, as I weave, the sea roils and boils with life. About a day and a half ago, the oceans were ailing from pollution. Today, as the sun rises, there may as well be a sign on all Lagos beaches that reads: 'Here There Be Monsters'. [...] I hope they do not understand any of it. If they do, then they will not step on to that boat and the story will not continue. [...]. I am Udide, the narrator, the story weaver, the Great Spider (Okorafor, 2014, p. 228).

A weaver spider, therefore, of spatial and narrative, temporal and regenerative plots, which take us back to Haraway's *Chthulucene* trouble—and thus to oceanic turbulence—whose weaver spider is precisely that *Pimoa Chthulu*, a FS figuration to speak of a world that is, was and could be through all possible tentacular alliances. The *Pimoa chthulu* thus allies itself with the creatures of the seas, such as the coral reefs born from the severed head of Medusa, octopuses, squids, cuttlefish, the Octopus *cyanea* and the Mediterranean ophiuroid: tentacular beings that help us plotting with the turbulent oceanic materiality—and thus with other ways of becoming-with, of building alliances, of imagining political ecologies, as occurs in *Lagoon*.

These sea creatures, sprawling, regener-a(c)tive, are key figurations for understanding and thinking with the *Chthulucene*: a world of matter and metaphor, of bodies and relationships, which find its site par excellence in the space of the sea and its turbulent materiality.

In the sprawling era of the *Pimoa Chthulu*, the possibility of cultivating imaginative collective practices and the construction of other worlds is still in its making, if we do not give up the ability to imagine, think and tell: 'In passion and action, detachment and attachment, this is what I call cultivating response-ability; that is also collective knowing and doing, an ecology of practices. Whether we asked for it or not, the pattern is in our hands. The answer to the trust of the held-out hand: think we must' (Haraway, 2016, p. 56).

4.4 Conclusions

In this chapter I tried to interweave together the Critical Ocean Geographies developed inside the Oceanic Turn (especially the more-than-wet geographies elaborated by Steinberg and Peters) with the Blue Humanities developed inside the feminist posthuman New Materialism. As an alliance of knowledges trying to clash against the separation of matter and meaning, these onto-epistemologies can help us through diffraction to build situated political speculative knowledges and stories against the catastrophism of Anthropocene and Capitalocene. The sea, in particular, is once again the pivotal space of conflicting vectors and possibilities that can suggest us the development of other relationship and collaborative ecologies. Its turbulent materiality challenges and repositions notions and relationships of space and time, made of flows, liquidity, other volumes, circularity, intra-actions, processes of trans-corporeality. As plotted in the Africanfuturistic novel *Lagoon* by Nnedi Okorafor, the sea is also the knot from which interweave, think, imagine, practice possible futures. But these futures need to be situated, as we are not all response-able in the same way and stories are not neutral: this is way Lagoon starts from the West Coast of Africa, where a new world outside the violence racial and extractive capitalism has to be rethought.

It is from the sea-space and from its turbulent materiality that we can start thinking otherwise, decentralising our gaze, situating violent stories, developing a world of response-ability that concerns living and dying well. And since we are bodies of water embedded in and connected to the ocean, these futures, these worlds, these alternative ecologies can only start from oceanic matter and its blue kinship, figurations, sprawling relationships, tentacular alliances. A trouble of a turbulent materiality, in other words, that disturb us, question us, stir us.

References

Alaimo, S. (2012). States of suspension: Trans-corporeality at Sea. *Interdisciplinary Studies in Literature and Environment, 19*(3), 476–493.

Anderson, J., & Peters, K. (Eds.). (2014). *Water worlds: Human geography of the ocean*. Ashgate.

Balzano, A., Ferrante, A. A, Timeto, F. (2022). Afterword. The words of kinship: From translating to making. In Clarke, A., & Haraway, D. (eds.), *Making Kin. Making kinship, not populations* (pp. 183–198). DeriveApprodi.

Barad, K. (2007). *Meeting the universe halfway*. Duke University Press, Durham.

Barad, K. (2014). Invertebrate Visions: Diffractions of the Brittlestar. In E. Kirksey (Ed.), *The Multispecies Salon* (pp. 221–241). Duke University Press.

Bennett, J. (2010). *Vibrant Matter: A Political Ecology of Things*. Duke University Press.

Borghi, L. (2017). Foreword. In *Barad K Performativity of nature. How much and queer* (pp. 6–9). ETS Editions.

Braidotti, R. (2013). *The posthuman*. Polity Press.

Brathwaite, K. E. (1975, September). Caribbean man in space and time. *Savacou 11/12*, 1–11.

Butler, J. (1990). Gender Trouble. *Feminism and the Subversion of Identity*. Routledge.

Carson, R. (1951). *The sea around us*. Oxford University Press.

Colliva, T. (2021). Nnedi Okorafor: Trajectories of an African futurism. *Ptolemy, 23*, 135–152.

DeLoughrey, E. (2016). The Oceanic Turn: Submarine futures of the Anthropocene. In Adamson, J., & Davis, M. (eds.), *Humanities for the Environment. Integrating Knowledge, forging new constellations of practice* (pp. 242–258). Routledge.

Giuliani, G. (2015). *Zombies, aliens and mutants*. Le Monnier.

Gumbs, A. P. (2021). *Undrowned: Black feminist lessons from Marine Mammals*. AK Press.

Haraway, D. (1991). *Simians, cyborgs, and women: The reinvention of nature*. Routledge.

Haraway, D. (2016). *Staying with the trouble. Making Kin in the Chthulucene*. Duke University Press.

Helmreich, S. (2009). Alien Ocean. *Anthropological Voyages in Microbial Seas*. University of California Press.

Jue, M. (2017). Intimate objectivity. On Nnedi Okorafor's Oceanic Afrofuturism. *Women's Studies Quarterly, 45*(1–2), 171–188.

Lehman, J., Johnson, E., & Steinberg, P. (2021). Turbulent Waters in Three *Parts. Theory and Event, 24*(1), 192–219.

Massey, D. (2005). *For space*. Sage.

Neimanis, A. (2019). Bodies of water. *Posthuman Feminist Phenomenology*. Bloomsbury.

Okorafor, N. (2014). *Lagoon*. Hodder & Stoughton.

Palermo, G. (2024). Fatti e fabule. *Geographies more-than-wet of the Black Mediterranean*. Franco Angeli.
Peters, K., Anderson, J., Davies, A., & Steinberg, P. (Eds.). (2022). *The routledge handbook of ocean space*. Routledge.
Peters, K., & Steinberg, P. (2019). The ocean in excess. Towards a more-than-wet ontology. *Dialogues in Human Geography, 9*(3), 293–307.
Phillips, J. (2018). WATER. Order and Offshore: The Territories of Deep-Water Oil Production. In Peters, K., Steinberg, P., & Stratford, E. (eds.), *Territory Beyond Terra* (pp. 51–67). Rowman & Littlefield.
Pinto, I. (2021). Storytelling Multispecie. Una pratica ecopolitica per la giustizia ambientale. *Etnografie del contemporaneo, 4*(4), 33–46.
Serrano, A. (2020). The Rise of the Afrofuturistic Novel: The Intersection of Science Fiction and African Environmentalism in Nnedi Okorafor's Lagoon. *Revista Hélice VI, 2*, 39–57.
Sharpe, C. (2016). *In the wake: On blackness and being*. Duke University Press.
Steinberg, P. (2001). *The social construction of the ocean*. Cambridge University Press.
Steinberg, P. (2022) Blue planet, black lives: Matter, memory and the temporalities of political geography. *Political Geography, 96*, 102524, 1–13.
Steinberg, P., & Peters, K. (2015). Wet ontologies, fluid spaces: Giving depth to volume through oceanic thinking. *Environment and Planning d: Society and Space, 33*(2), 247–264.
Steinberg, P., & Palermo, G. (forthcoming). *The Trans-actional Ocean: Excessive Mobilities and Immersive Temporalities at Sea*. Social and Cultural Geography.
Timeto, F. (2020). Bestiario Haraway. Per un femminismo multispecie. Mimesis.
Vallorani, N. (2017). Nnedi Okorafor and the circles in the water. In Okorafor, N., Laguna. Zone 42, Modena, pp. 295–302.
Zurita, M., Munro, P. G., & Houston, D. (2018). Un-earthing the Subterranean Anthropocene. *Area, 50*, 298–305.

CHAPTER 5

Plastisphere in Chthulucene. "Staying with the Problem" in Oceanic Natureculture Assemblages

Chiara Certomà and Federico Fornaro

5.1 Introduction: Re-Shaping and Re-Signifying the Ecological Life-Supporting Systems

The escalating environmental problems of the Anthropocene, particularly concerning the global ocean, necessitate urgent and comprehensive action. These problems, which are already affecting the global ecosystem (EC, 2020), include the unsustainable exploitation of marine resources, the consequences of human-induced climate change and the marine pollution deriving from various sources (such as plastics, microplastics, chemicals and noise). Addressing these interconnected issues requires

C. Certomà (✉)
Sapienza University of Rome, Rome, Italy
e-mail: chiara.certoma@uniroma1.it

F. Fornaro
Raw-News Visual Production Agency, Epping, UK
e-mail: federico.fornaro@raw-news.net

C. Certomà (ed.), *Blue Kinships*,
https://doi.org/10.1007/978-3-031-78619-8_5

concerted efforts, beyond documenting changes, to actively bridging the gap between humanity and the Ocean.

Central to this endeavour is fostering a closer emotional connection to the sea, as already suggested by the U.N. "One Ocean" blueprint and the European Commission's emphasis on ocean literacy (EC, 2022; UNDP, 2022). Following the call of the ecofeminist postmodern philosopher Donna Haraway tightening society's bond with the ocean calls for increasing and improving the number and quality of the relationships between humans and non-humans alike in a timespace she calls Chthlucene (Haraway, 2015), when we are pressed to live and deal with the troubles generated by anthropic activities. Haraway's invitation to "stay with the trouble" entails the adoption a multispecies perspective to address the complexities of the Anthropocene; and "demand[s] our best emotional, intellectual, artistic, and political creativity, individually and collectively" (Haraway, 2015, p. 164).

The emergence of the plastisphere, comprising organisms that colonise plastic in marine environments, represents a material and symbolic manifestation of the Chthulucene natureculture assemblages. These assemblages' essence, composition and life challenge our understanding of the traditional ontological and epistemological boundaries and call for reconsidering oceanic kinships based on sympoietic processes. We approach here the new hybrid assemblages that are re-shaping and re-signifying the ecological life-supporting systems by building upon the material-semiotic of more-than-human theory, to cope with the environmental transformations in an age characterised by subterraneous connections.

To this end, we embrace the sympoietic process as the chief process forging new kins between societies and the sea; and, by drawing on the compositions of materialities, imaginaries and forces in the liquid marine space, we knit again our common belonging to the ocean. This corresponds to the exercise of a "tentacular thinking", a perspective emphasising the tentacular nature of a life lived along interconnected lines and interlaced trails that make biology appear as a process of constant change and transformation, to forge intra-species kinships and relational values connecting society and the ocean. To engage with the sympoietic practice we need to move beyond the purification and separation practices of modern epistemology to recognise the existence of porous boundaries and fluid mutable connections between living and non-living matter.

5.2 Plastisphere, Emerging Ecological Formations in the Global Ocean

Plastic pollution poses a significant threat to marine ecosystems, with about 20 million metric tonnes of plastic entering aquatic environments annually (Borrelle et al., 2020). Despite global efforts to combat plastic pollution, it is expected to worsen, leading to far-reaching environmental, economic and social consequences (Beaumont et al., 2019; Schnurr et al., 2018; Xanthos & Walker, 2017). While the Great Pacific garbage patch is the most well-known concentration of seaborne plastic waste, plastic pollution is pervasive in all oceanic regions; and it is significantly affecting marine environment and marine life physically, physiologically and behaviourally (Gall & Thompson, 2015; Green et al., 2019). Research is uncovering new forms of life and adaptation of existing ones to the changed marine environment. For instance, Jagiello et al. (2024) reported on hermit crabs exhibiting novel behaviour related to the use of plastic as protective shells, highlighting the widespread availability of artificial shells due to plastic pollution (Picture 5.1).

Plastic waste (polyethene, polypropylene and expanded polystyrene) in marine contexts are often identified with macroplastic, solid plastic objects clearly visible almost everywhere and in every season, whose being "matter out of place" (Douglas, 1979) is iconically represented by images of sea inhabitants caught in plastic nets or bags. However, macroplastics fragmentate in floating particles that end up below the ocean surface in different oceanic compartments via biofouling and enter the marine food web via ingestion. The differentiation between microplastics and macroplastics carries significant implications for the conceptualization and definition of pollution, particularly within the domain of marine litter where this categorization originated (Bergmann, 2022). Microplastic, i.e. particles of plastic smaller than five millimetres in diameter, are even more pervasive than macroplastics in the marine environment. These particles can absorb and transport toxic chemicals, threatening marine biodiversity and ecosystem health (Bergmann, 2022). Microplastics form intricate associations with plankton and other organic particles within marine snow, which includes various living and non-living particles such as copepod excreta, and, while relatively inconspicuous on the ocean surface, they pose difficulty in differentiating waste from organic matter. These blur traditional distinctions between synthetic and natural elements and complicating the categorisation of waste versus life (De Wolff, 2017).

Picture 5.1 Spring water. Plastic bottle on rocky substrate with corraligenous formations (photo Giuseppe Lupinacci)

Marine plastic hosting simple colonising organism or complex communities, in fact, represent unique ecosystems known as "plastisphere". The term plastisphere was first coined by Zettler et al. in 2013 to refer to a collective of plastic-colonising organisms, including bacteria and fungi triving on a biofilm that provides the necessary chemical support. The bacterial appetite for plastic is motivated by the concentration and accessibility of nutrient like nitrogen, iron and phosphorus on plastic surfaces, more than in depleted oceanic areas (Zettler et al., 2013).

Distinct microbial communities with potential and mostly unknown implications for ecosystem function and disffusion are being mapped in the plastisphere (Amaral-Zettler et al., 2020; Li et al., 2021), where "there are organisms that photosynthesise; there are predators and prey; symbionts and parasites" (Thomas, 2021b). Initially, research on the plastisphere primarily utilised microscopy to distinguish organisms like diatoms and filamentous bacteria, most of which naturally form biofilms and prefer an attached lifestyle. With molecular techniques, particularly high-throughput DNA sequencing, today our comprehension of the diverse microorganisms inhabiting the plastisphere is expanding. This indicates differences in the microbial communities on plastic compared to other particles in the same environment, with certain microbial groups consistently present on plastic—despite, as far as we know, there are no microorganisms exclusive to plastic. Nevertheless, microorganisms are expected to soon adapt in colonising novel plastic substrates.

Since its first identification, the reference of the term plastisphere expanded to loosely encompasses larger organisms, from crabs to jelly-fish, which raft across oceans on marine plastic debris or with plastic ingested. In fact, unlike most natural materials, plastic is highly durable, allowing the growth and spread of attached organisms over a massive area or allowing its persist in living bodies for a time longer than the life of the organism itself. As Haram et al (2023) note: "Rafting, or the association of organisms with floating debris, has been an inferred mode of marine species dispersal. Enduring plastic materials may survive much longer, although degradation rates vary across polymer type, habitat and environmental conditions. Transoceanic dispersal on rafts is a high-risk, low-probability endeavour for coastal organisms, such that oceans typically represent. Our results suggest that the historical lack of available substrate limited the colonization of the open ocean by coastal species, rather than physiological or ecological constraints as previously assumed. It appears that coastal species persist now in the open ocean as a substantial component of a neopelagic community sustained by the vast and expanding sea of plastic debris" (Haram et al., 2023, p. 687). Through rafting bacterial population disperse in open sea and colonise new areas of the global ocean, making plastic a vector for a large number of microbes including pathogenic microbes, microbes showing antibiotic resistance, and those that show a capacity to compost plastic. This evidence brought Robyn Wright (Thomas, 2021b) to note that: "I don't think it's necessarily important that it's not natural in origin, because all of the members of the plastisphere are still 'natural', but it's more an issue of scale" of the problem. From the one side the proved existence of plastic-consuming organisms generate hope on the potential of bioremediation strategies (such as harnessing plastic-degrading bacteria), but from the other side this raises concerns about unintended consequences (Yoshida et al., 2016) and the potential exacerbation of the problem by relying on easily solutionist approach (Bergmann, 2022).

5.3 Oceanic Assemblages. Follow the Things into the Deep Blue

Acknowledging the pervasive infiltration of plastic into the global biochemical system, biologist Cristina Romero Castillo highlights (2022) the enduring impact of plastic pollution. In describing the human footprint on the ocean, she named it as AnthropOceano and noted that "Many people have proposed projects to clean up the ocean of plastic, for example. But that's like sweeping in the desert. Most of the plastic that has been reaching the sea over the years has been broken into smaller and smaller pieces, better known as microplastics, and that is impossible to clean up. Now we can only wait for those little pieces to end up sedimenting and end up buried in the seabed, forming a layer that in the future will be a witness of this era of 'plasticene'" (ICM, 2022). Its removal is virtually impossible; and furthermore, it is questioned whether it would be desirable, considering the vitality and entanglement of plastics with living organisms (Scott, 2018) (Picture 5.2).

Picture 5.2 Floating plastic. Female pad and plastic snow floating with algae (photo Giuseppe Lupinacci)

The massive presence of plastic in the sea opens up ontological, epistemological and pragmatic questions. Notably, the plastisphere urges us to interrogate the ontology and meaning of new marine ecologies by revising the significance of the classifications adopted (Bowker & Starm, 2000). To this end, we need to understand and deconstruct the modern conceptual categorisation as the result of a constant work of purification operated by the scientific research. Anthropologist Mary Douglas argued in Purity and Danger (1966) that since ancient times purity was referred to as the absence of contamination, of pollution, not only in an ecological sense but in a historical, social or cultural sense as well. She explained that in this perspective "Danger lies in the transactional state because transition is neither one state nor the next, it is indefinable" (p. 96). Pollution is a particular class of danger associated with ambiguity that emerges from the absence of control and definition, and it "is a type of danger which is not likely to occur, except where the lines of structure, cosmic or social, are not clearly defined" (p. 113). Pollution and untidiness has been referred to the subversion of the ontological order when this contemplates the existence of several causal relations among all the ontological levels of being. Building upon Douglas's research, Bruno Latour argued in We have never been modern (1993) that our scientific and technological culture avoids ontological "pollution" by separating humans and non-humans, purified entities and hybrids mixtures. The (supposed) power of modern science resides precisely in its ability to maintain the distinctions by hiding the hybridisation that constantly occour in daily practices and by "conceiving every hybrid as a mixture of two pure forms" (Latour, 1993, p. 78). Therefore, modern science and culture prioritise the practices of purification that allow us to categorise and clearly distinguish everything existing into sharply separated domains. The need for purification roots in the moderns' obsession for hybrids that "present the horror that must be avoided at all cost by a ceaseless, even maniacal purification" (Latour, 1993, p. 37), despite laboratories and factories every day populate the world with thousands of hybrids and invent entirely new matters.

The attempt of prioritising purification and hiding hybridization has been accomplished, Latour writes: "[b]y rendering mixture unthinkable, by emptying, sweeping, cleaning and purifying the [public] arena [...], the moderns allowed the practice of mediation to recombine all possible monsters without letting them have any effect on the social fabric, or even any contact with it. Bizarre as these monsters may be they posed no problem because they did not exist publicly and because their monstrous consequences remained untraceable" (Latour, 1993, p. 42). As a matter of fact, the "monstruos consequences" of hybrids of nature and culture (which Donna Harawy names as the cyborgs, the tricksters, the monsters) are evidently manifesting in the plastisphere. Categories, separations, orders and hierarchies are cultural constructions; they refer to the epistemological, moral and political spheres, not the ontological one.

In her article titled "Speculating with Organisms in the Plastisphere", Jennifer Gabrys (2014, p. 57) posits that bacteria possess not only the capacity to sustain themselves but also to modify their surroundings, effectively transforming into a different organism in the process. She suggests that this transformative ability is inherent in the speculative nature of organismal life. The introduction of plastic into marine environments resembles an experiment with unforeseeable outcomes, Gabrys contests, whose consequences include the rise to novel life forms in the ocean. The emergence of the plastisphere is making natureculture "monstrous", uncanny assemblages visible and not hidable anymore; and through the plastisphere, natureculture assemblages are materially transforming and culturally re-signifying the global marine life-supporting systems and "problematize the current ways in which plastic is described not only as litter but also as pollution".

Considering these considerations, Bergmann (2019) prompts a critical inquiry: what alternative frameworks can we devise to transcend conventional divisions between nature and culture, particularly when confronting modern challenges like climate change and marine debris? Disentangling the plastisphere's multiple materialities and polysemic nature requires the adoption of a material-semiotic perspective, which conceives "every in the social and natural worlds as a continuously generated effect of the webs of relations within which they are located" (Law, 2007, pr. 1.). The heterogeneity of the emergent marine ecosystems offers a lockpick to debunk the power of binary, dualist and oppositional thinking, by conceiving the spatial and material-semiotic networks we are all embedded within as constitutive part of what is "stitched together' across divisions and distinctions" (Murdoch, 2006, p. 322). Nowhere as in the liquid oceanic environment the plastisphere gives substance to theoretical speculations. Here borders are porous, there are gradual forms of hybridization that originate completely new entities, which recalls Steven Hinchliffe's definition of natureculture assemblages as "more or less durable bodies made up of similarly hybrid and impermanent relations. Things are [...] configured, or drawn together, in order to become more or less stable forms. There are no pre-existing essences, only relations" (Hinchliffe, 2007, p. 80).

As posited by Bergmann (2019), marine scientists navigate through contrasting conceptualizations of nature, oscillating between an interpretation wherein human agency seemingly exerts profound influence, even to the extent of nuclear contamination potentially fostering novel life forms, and an alternate view of nature as imperilled and necessitating protective measures against anthropogenic intervention. This dichotomy underscores the complexity inherent in contemporary ecological discourse. The notion of entangled "assemblages" elucidates the intricate interconnection between human actors, other biological entities and the environment. Interaction, interference or co-creation of hybrids are the result of co-dependence, which enacts an increasing differentiation because natureculture assemblages are not constituted once for all, but are in fiery entities (Whatmore, 2002). While in the realm of marine debris science, entanglement is typically viewed negatively as a state of non-belonging, from a material-semiotic perspective they represent an occasion to challenge established boundaries between plastic and animal, synthetic and natural, waste and life.

For instance, as part of the material-semiotic methodological apparel, the "follow the thing" methods (Cook, 2017) describes the interconnecting between human and non-human agents, thereby localising and examining how these entities actively shape meanings and conditions through performative actions (Latour, 2004), and explores how these entities dynamically alter conditions and events. Hence, this approach transcends mere methodological inquiry, advocating for the mapping of networks or assemblages comprising entities, their meanings, agencies and values, their connections and disconnections with other entities and actors, thereby prompting geographers to reconsider the socio-economic, cultural, political and environmental ramifications engendered by such network. This methods would allow us to appreciate the submarine entanglements into the deep blue, to unveil the intimal chemico-physical bounds that cannot be unbounded. In this way, the fate of a plastic bottles, produced and shipped along global trade networks to the retail market, only to be discarded and end up in the sea where its fragment into particles, can be followed up to when it become home to a variety of microorganisms. By tracking the journey of plastic in the marine environment, from its macroplastic to microplastic dimensions, scrutinising its colonisation by bacterial biofilms that prepare the substrate for more complex organism colonisation, and tracing the drifting of plastic microcosms through the ocean to their arrival in new niches, and finally, their integration into global biochemical systems, we can make a case for practising material semiotics in the wet and more-than-wet marine environment (SteinbergPeters, 2015).

5.4 Impossible Disentanglements

The complex interrelationships within the plastisphere cannot be easily disentangled, and this suggests a need for nuanced approaches to understanding and managing marine plastic. The persistence of these plastic entanglements underscores the interconnectedness between humans and the ocean, challenging us to reconsider our role in shaping and understanding marine environments. Thus, plastic's integration or exclusion from these networks should be understood within the broader context of its relationships, wherein humans are inexorably intertwined. Sorting practices and classification systems underlie political implications and are performative in nature. Plastic and life coexist and coalesce, as evidenced by studies on rafting (Thomas, 2021b), where marine biologists and ecologists investigate how species distribution is influenced by the presence of plastic in their environments. Marine organisms find themselves entangled with plastic, altering their relationships with their surroundings. Plastic, in turn, becomes intertwined with life, serving as rafts for organisms such as jellyfish and acting as carriers for small plastic fragments ingested by birds and fish. By accepting rather than attempting to untangle existing entanglements, synthetic materials can be comprehended as integral components of complex ecological networks (Picture 5.3).

The plastisphere challenges the conventional understanding of pollution by revealing relationships that actively support life rather than merely intersecting with it. In a multispecies ethnographic study focusing on plastic naturecultures, Kim De Wolff (2017) traced the movements of plastic creatures within the context of public education campaigns. This investigation specifically scrutinised the underlying assumptions regarding belonging and agency in the discourse surrounding whether and when plastic species should interact. De Wolff examined the activities of volunteers engaged in the meticulous sorting of minuscule plastic particles

Picture 5.3 Taking roots on the plastisphere. Alcyonium palmatum grown up on a plastic bottle caught in fishermen's net (photo Giuseppe Lupinacci)

from organic matter under a microscope. Through this process, volunteers established material demarcations, determining what constitutes life (non-plastic) and what does not (plastic), being the latter portrayed as 'dangerous' and supporting 'non-native species'. Nevertheless, DeWolff expands the analysis of plastic-life relationships in marine environments beyond the conventional narrative of 'harmful' plastic threatening an 'untouched' ocean 'ecosystem'. By adopting a perspective that acknowledges the unavoidability and inextricability of oceanic naturecultures entanglements, she illustrates how the conceptual categories of natural and artificial, along with assumptions regarding the feasibility and necessity of removing plastic from the sea, emerge from a dichotomic understanding of life and plastic in marine ecosystems. Consequently, the study delved into the inherent indeterminacy—or plasticity—of the classifications. De Wolff states "I argue that the 'danger' of plastic relationships lurks not in associations but in the very categories used to know and live with forms of plastic and forms of life, in the kinds of belonging that emerge with kinds of materials, and in the failure to recognise the impossibility of their separation. I learned that the plastisphere is made of the 'associations' and 'novel interactions' of life forms defined by their relationship to a human product in the ocean: eggs laid on a plastic bag, gooseneck barnacles burrowed into foam, bacteria nestled on synthetic fragments [...]. Plastic in the ocean is not simply human-made 'matter out of place' [...]; it is also a matter of life (and death) as bodies come to live, travel and die with plastic, and with plastic inside them" (De Wolff, 2017, p. 24). The perceived danger of plastic-life relationships does not merely lye in the associations themselves but in the frameworks used to understand and interact with different materials, emphasising the impossibility of their complete separation. Therefore, DeWolff (2017) proposes transcending the attempt of purification by plastic removal from the sea because, the underlying assumptions of these clean-up endeavours prioritise notions of belonging, assign value judgments to plastic, and impose human responsibility for environmental degradation, disregarding the communities intertwined with ocean pollution. Removing waste from the water may not necessarily benefit the organisms it shelters, and it might determine unintended consequences, such as leaving coastal reef fish vulnerable to predation in the open ocean when plastic shelters are removed. Nonetheless, most of the prevailing narrative continues to reinforce the separation between nature and culture, overlooking the intricate interconnections that define the plastisphere. The complex entanglements comprising the plastisphere defy simple purification practices, although scientific and public approaches to addressing plastic pollution in the ocean tend to perpetuate the dichotomy between nature and culture. Instead of dismissing the complexities of the plastisphere, DeWolff invites us to embrace the plastisphere science, which acknowledges the dynamic interplay of bodily associations and relational dynamics (2017).

Fom a pragmatic perspective, Bergmann clarifies "A natureculture perspective [...] calls into question existing notions of what belongs and what does not belong in a particular environment" (2019, p. 1). The messy business of living together leads to a relational ontology acted by heterogeneous associations, whose agency in marine environment has been already analysed in marine political ecology research (see Bennet, 2019). For instance, a place is not only created in the common sense by administration' and citizens' will, but it is also claimed by the sea, invaded by the birds, transformed by the production plant waste, flooded by the rains, connected by wireless internet connection...all those are nothing less than effective forms of place-making, of imposing or proposing a particular view of what a place should be, of showing a way of dwelling it. For instance, in his examination of the controversy surrounding scallop dredging in Cardigan Bay, Wales, Bear (2012) depicts the "scallop fishery" as a complex arrangement of diverse elements, both gathering and dispersing, and intertwining multiple temporal and spatial dimensions through processes of territorialisation and deterritorialisation to define what are we restoring, protecting and enhancing through our interventions. This emphasizes also the entwined relationship between living and non-living elements, urging a re-evaluation of their significance and the formulation of strategies for accountability and care.

5.5 Staying with the Plastisphere Trouble?

If we depart from purification-focused strategies, we can better understand the intrinsic vitality and interconnection between plastic and various organisms. We can consider the plastics in the Ocean as an unintentional experiment bringing potential transformations of the environment and necessitating new conceptualisations for problematising and addressing it pragmatically because the pervasiveness of plastic pollution makes the blurring of physical and conceptual boundaries between natural and synthetic environments apparent.

Still acknowledging the impossible disentanglements of plastic and organic matter, Bergmann (2016) stresses the importance of not remaining politically indifferent to the issue of plastic pollution. Understanding the complex dynamics between plastics and oceanic biota is essential to define what interventions are to be undertaken, considering that plastic is not merely a pollutant but an integral component of emerging habitats. The intricate web of linkages, entanglements and enmeshments complicates efforts to address marine plastic pollution solely through clean-up initiatives, as these interventions risk disrupting the interwoven fabric of ecosystems. The notion of purification, therefore, proves not only conceptually flawed but also practically challenging, revealing the inadequacy of simplistic approaches to cleansing the oceans of plastic contaminants. For sure, recognizing the value of entanglements prompts a reassessment of intervention strategies, advocating for a more comprehensive approach to the prevailing capitalist production paradigm relying on the relentless global production of plastics, which often surpasses genuine societal needs.

Notably, acknowledging the need to coexist with emergent and entwined assemblages of humans, non-humans and more-than-humans within the oceanic environment provides a foundation for critically analysing our relationship with the ocean in our present time Haraway calls the Chthulucene, "a kind of timeplace for learning to stay with the trouble of living and dying in response-ability on a damaged earth" (Haraway, 2016). In the Chtulucene biology manifests as a dynamic process of continual transformation and "posthuman becoming", wherein all organisms are co-evolved symbionts (p. 102). The Chthulucene is named after the chthonic creatures that "demonstrate and perform the material meaningfulness of earth processes and critters" (Haraway, 2016, p. 19), that create linkages, attachments, ongoing multistories and make kin which is "oddkin rather than, or at least in addition to, godkin and genealogical and biogenetic family troubles important matters, like to whom one is actually responsible" (p. 19). Their existence has been systematically symbolically and materially denied by the technocratic solutionism proposed in the Capitalocene and the Anthropocene. To dismiss the faith in technocratic fixes or the fatalistic resignation, Haraway advocates for "staying with the trouble", and emphasises our present engagement with the myriad unfinished configurations of existence (Haraway, 2016, p. 18).

Her conceptualisation of the Chthulucene urges us to reassess our interactions with non-living entities, including potentially harmful plastics, to coexist on a damaged Earth where the human and non-human are inextricably linked. For instance, she emphasizes the significance of coral reefs, serving as vibrant ecosystems intertwined with the well-being of millions of humans. The flourishing of both diverse corals and human populations is contingent upon a multispecies response-ability, lest the delicate balance of biodiverse ecosystems collapse under strain. She explains that coral reefs have played a pivotal role in highlighting the impacts of human activities, particularly the Anthropocene's effects such as ocean warming and acidification due to CO2 emissions. These stressors detrimentally affect coral health, leading to bleaching events and the loss of symbiotic relationships crucial for reef ecosystems. Moreover, corals and lichens also recall the exploitative practices of the Capitalocene era, like deep-sea mining and fracking, which disrupt delicate ecosystems and exacerbate global environmental challenges. Yet, within the narrative of the Chthulucene, corals and lichens offer a narrative of hope, inviting collaborative efforts towards a sustainable future, embracing symbiotic relationships and metabolic transformations for collective well-being.

In alignment with Haraway's inquiry, we contemplate the multifaceted impacts of biocultural, biotechnical and biopolitical factors on historically situated individuals, recognising the interconnectedness of human and non-human species assemblages within the context of evolutionary and socio-cultural narratives (Haraway, 2016, p. 117). This interconnectedness fosters the emergence of "mutant ecologies", exemplified by microbial colonisation of marine plastic, reflecting a dynamic process of "becoming-with" (Bergmann, 2019). The process of biochemical co-composition interrogates us about our shared belongingness to the oceanic realm resulting from the sympoietic process, involving materialities, imaginaries and forces. As clarified by Bergman (2016), sympoiesis involves the intricate interpenetration and co-creation of critters, resulting in the formation of cells, organisms and ecological assemblages. Therefore, the plastisphere offers us a prominent example of the new temporal-spatial context we inhabit, underscoring the role of sympoiesis in shaping earthly ecologies, particularly oceanic systems. Moreover, the plastisphere phenomenon urges us to adopt a biopolitical perspective, considering the intricate interplay between synthetic materials and ecological systems in shaping socio-political narratives and imaginaries. An inclusive and integrated approach to oceanic stewardship (Jefferson et al., 2015; Peters et al., 2022) signals the need to reconsider both narrative constructs and pragmatic conditions shaping the composition of marine life amidst plastic pollution. This understanding can inform geographical research and facilitate the development of relational values that promote behavioural change towards marine conservation (Mckinley et al., 2020).

This could be possible because the sympoietic process enables us to exercise a "tentacular thinking", emphasising the tentacular nature of life lived along interconnected lines and interlaced trails, a perspective that presents biology as a process of constant change and transformation, forging intra-species kinships and relational values. Interpreted as a form of reciprocity and responsibility with other living species and non-living matter, all depending on the ocean and its health, tentacular thinking allows us to cultivate a closer relationship between society and the ocean. This way we can grasp links, and deep affinities, overcome distances, and appreciate the multitude of hybrid and heterogeneous connections bounding all earthly beings and things together.

References

Amaral-Zettler, L. A., Zettler, E. R., & Mincer, T. J. (2020). Ecology of the plastisphere. *Nature Reviews Microbiology, 18*(3), 139–151.

Bear, C. (2012). Assembling the sea: Materiality, movement and regulatory practices in the Cardigan Bay Scallop Fishery. *Cultural Geographies, 20*(1), 21–24.

Beaumont, N. J., Aanesen, M., Austen, M. C., Börger, T., Clark, J. R., Cole, M., et al. (2019). Global ecological, social and economic impacts of marine plastic. *Marine Pollution Bulletin, 142*, 189–195.

Bennett, N. J. (2019). Political seas: Engaging with political ecology in the ocean and coastal environment. *Coastal Management, 47*(1).

Bergmann, S. (2019). Schleimige Assoziationen im Meer—die Plastisphäre. In F. Gesing, M. Knecht, M. Flitner, & K. Amelang (Eds.), NaturenKulturen. Denkräume und Werkzeuge für neue politische Ökologien (pp. 353–384). Bielefeld: transcript.

Bergmann, S. (2022). Dawn of the plastisphere: An experiment with unpredictable effects. In S. Bergmann (Ed.), *Plastic legacies: Pollution, persistence, politics*. Athabasca University Press.

Borrelle, S. B., Ringma, J., Law, K. L., Monnahan, C. C., Lebreton, L., McGivern, A., et al. (2020). Global plastic waste generation exceeds efforts to mitigate plastic pollution. *Science, 369*(6510), 1515–1518.

Bowker, G. C., & Star, S. L. (2000). *Sorting things out: Classification and its consequences*. MIT Press.

Cook, I. et al. (2017). From 'follow the thing: Papaya' to followthethings.com. *Journal of Consumer Ethics, 1*(1), 22–29.

De Wolff, K. (2017). Plastic naturecultures: Multispecies ethnography and the dangers of separating living from nonliving bodies. *Body and Society, 23*(3), 23–47.

Douglas, M. (1979). *Pureness and danger. An analysis of concepts of pollution and taboo.* Routledge & Kegan Paul.

European Commission. (2020). Directorate-General for Research and Innovation. In Lamy, P., Citores, A., Deidun, A. et al. (Ed.), *Mission Starfish 2030—Restore our ocean and waters*. Publications Office. Retrieved from https://research-and-innovation.ec.europa.eu/knowledge-publications-tools-and-data/publications/all-publications/mission-starfish-2030-restore-our-ocean-and-waters_en

European Commission. (2022). One ocean summit: New steps strengthen EU leadership in protecting the Ocean, Brest, France, 11 February 2022. Retrieved from https://knowledge4policy.ec.europa.eu/news/one-ocean-summit-new-steps-strengthen-eu-leadership-protecting-ocean_en

Gabrys, J. (2014). Speculating with organisms in the plastisphere. In P. Yoldas (Ed.), *An ecosystem of excess* (pp. 48–59). Ernst Schering Foundation.

Gall, S. C., & Thompson, R. C. (2015). The impact of debris on marine life. *Marine Pollution Bulletin, 92*(1–2), 170–179.

Green, D. S., Colgan, T. J., Thompson, R. C., & Carolan, J. C. (2019). Exposure to microplastics reduces attachment strength and alters the haemolymph proteome of blue mussels (Mytilus edulis). *Environmental Pollution, 246*(March), 423–434.

Haram, L. E., Carlton, J. T., Centurioni, L., et al. (2023). Extent and reproduction of coastal species on plastic debris in the North Pacific Subtropical Gyre. *Nature Ecology & Evolution, 7*, 687–697.

Haraway, D. (2015). Anthropocene, Capitalocene, Plantationocene, Chthulucene: Making Kin. *Environmental Humanities, 6*(1), 159–165.

Haraway, D. (2016). *Staying with the problem: Making Kin in the Chthlucene.* Duke University Press.

Hinchliffe, S. (2007). *Geography of nature: Societies, environments, ecologies.* Sage.

ICM. (2022). Cristina Romera-Castillo: "Cleaning the ocean out of plastic is like sweeping in the desert".

Jagiello, Z., Dylewski, Ł., & Szulkin, M. (2024). The plastic homes of hermit crabs in the Anthropocene. *Science of the Total Environment, 913.*

Jefferson, R., et al. (2015). Understanding audiences: Making public perceptions research matter to marine conservation. *Ocean & Coastal Management, 115.*

Latour, B. (1993). *We have never been modern.* Harvard University Press.

Latour, B. (2004). *Politics of nature.* Harvard University Press, Harvard.

Law, J. (2007). Actor network theory and material semiotics. Centre for Science Studies, Lancaster University. Retrieved from [http://www.heterogeneities.net/publications/Law2007ANTandMaterialSemiotics.pdf]

Li, C., Wang, L., Ji, S., Chang, M., Wang, L., Gan, Y., & Liu, J. (2021). The ecology of the plastisphere: Microbial composition, function, assembly, and network in the freshwater and seawater ecosystems. *Water Research, 202.*

McKinley, E., Acott, K., & Yates, K. (2020). Marine social sciences: Looking towards a sustainable future. *Environmental Science & Policy, 108.*

Murdoch, J. (2006). *Post-structuralist geography.* Sage.

Peters, K., et al. (Eds.) (2022). The routledge handbook of ocean space. Taylor & Francis.

Romero Castillo, C. (2022). Antropoceano. Cuidar los mares para salvar la vida. Espasa.

Schnurr, R. E., Alboiu, V., Chaudhary, M., Corbett, R. A., Quanz, M. E., Sankar, K., et al. (2018). Reducing marine pollution from single-use plastics (SUPs): A review. *Marine Pollution Bulletin, 137*, 157–171.

Scott, E. (2018). Excerpts from select papers written by UO graduate and undergraduate students responding to works in the exhibition Art History 410/510: Land and Environmental Art | Fall 2018.

Steinberg, P., & Peters, K. (2015). Wet ontologies, fluid space. *Environment and Planning D, 33*(2), 247–264.

Thomas R. (2021a). Plastic rafting: The invasive species hitching a ride on ocean litter. *The Guardian*.

Thomas R. (2021a). Welcome to the 'plastisphere': The synthetic ecosystem evolving at sea. *The Guardian*. https://www.theguardian.com/environment/2021/jun/14/plastic-rafting-the-invasive-species-hitching-a-ride-on-ocean-litter

UNDP. (2022). *One ocean*. Retrieved from https://oneocean.undp.org/

Whatmore, S. (2002). *Hybrid geography*. Sage.

Xanthos, D., & Walker, T. R. (2017). International policies to reduce plastic marine pollution from single-use plastics (plastic bags and microbeads): A review. *Marine Pollution Bulletin, 118*(1–2), 17–26.

Yoshida, S., Hiraga, K., Takehana, T., Taniguchi, I., Yamaji, H., Maeda, Y., et al. (2016). A bacterium that degrades and assimilates poly(ethyleneterephthalate). *Science, 351*, 1196–1199.

Zettler, E. R., Mincer, T. J., & Amaral-Zettler, L. A. (2013). Life in the "plastisphere": Microbial communities on plastic marine debris. *Environmental Science & Technology, 47*(13), 7137–7146.

CHAPTER 6

An Intimate Connection: Society and the Ocean in Water Photography

Giuseppe Lupinacci

Interlude

G. Lupinacci (✉)
Raw-News Visual Production Agency, Epping, UK
e-mail: g.lupinacciphoto@gmail.com

C. Certomà (ed.), *Blue Kinships*,
https://doi.org/10.1007/978-3-031-78619-8_6

Picture 6.1 Oceanic society

Picture 6.2 Intestistial space: view of land from the tube

Picture 6.3 The imagined sea

Picture 6.4 Thirsty ocean

Picture 6.5 The word is reflected in water

Picture 6.6 Water without time. Historical ruins on the coast

Picture 6.7 Life on water. Forging the coastal shape

Picture 6.8 Meeting in water, meeting with water

Picture 6.9 My submarine garden

Picture 6.10 The sound of the ocean

Picture 6.11 Inside. In search of an intimate relationship

Picture 6.12 The sacrality of water. Celebrating the sea

Picture 6.13 Living of the sea. Fishermen

Picture 6.14 Riding the wind

Picture 6.15 Navigation. The ancient art of crossing ocean

Picture 6.16 Tightening watery relationship

Picture 6.17 Into the blue. Diving at minus 45 meters

PART II

Making the Ocean

CHAPTER 7

Citizen Science and Ocean Literacy: Bridging the Gap Between Society and the Ocean

Stephen Parkinson, Sasha Marie Woods, James Sprinks, Bianca Cruz, Kaushiki Das, Anna Witter, Gennadi Lessin, Dori Edelist, Dror Angel, and Luigi Ceccaroni

7.1 Introduction

The contribution of citizen science to marine management is increasingly being recognised and the field of marine citizen science is growing (Garcia-Soto et al., 2021; Wehn et al., 2025). Citizen science can be understood as "work undertaken by civic educators together with citizen communities to advance science, foster a broad scientific mentality, and/or encourage democratic engagement, which allows society to deal rationally with complex modern problems" (Ceccaroni et al., 2017, p. 1). The value of marine citizen science is often seen in the data and research

S. Parkinson (✉) · S. M. Woods · J. Sprinks · B. Cruz · K. Das · A. Witter · L. Ceccaroni
Earthwatch, Oxford, UK
e-mail: sparkinson@earthwatch.org.uk

S. Parkinson
University of Nottingham, Nottingham, UK

G. Lessin
Plymouth Marine Laboratory, Plymouth, UK

C. Certomà (ed.), *Blue Kinships*,
https://doi.org/10.1007/978-3-031-78619-8_7

insights it can provide, but citizen science projects also have an important cultural value and can provide numerous benefits to citizen scientists including capacity-strengthening and fostering a sense of emotional connection with the ocean (Sandahl & Tøttrup, 2020). Over the last two decades, ocean literacy has emerged as a term to describe a person's perception of the marine environment and can be defined as: an individual's understanding of the ocean's influence on them—and their influence on the ocean (Cava et al., 2005). Although the term was coined by American educators and emerged in an education context, the term has since been adopted world-wide and adapted to reflect a broader scope encompassing all societal groups, not just school pupils (MacNeil et al., 2021). The term has also started to broaden out to describe societal outcomes rather than just changes within individuals. For instance, the UN Ocean Decade Vision 2030 White Papers refer to "societal-level ocean literacy" (Glithero et al., 2024, p. 8). Ocean literacy can therefore be understood as a measure of the extent to which society feels it is connected to the ocean. Reflecting how the scope of ocean literacy has evolved and grown over time; ocean literacy was recently (re)defined in a framework of ten dimensions—see Table 7.1 (McKinley et al., 2023).

This chapter outlines the overlapping aims and impacts between ocean literacy and citizen science. In doing so, it aims to demonstrate the value of citizen science as a tool to bridge the gap between society and the ocean. First, the connection between citizen science and ocean literacy is established, drawing on evidence from a series of narrative literature reviews. Three projects are then presented as case studies (ProBleu, Iliad and FreshWater Watch) to further explore the connection between citizen science and ocean literacy. The chapter concludes with recommendations to increase the impact that citizen science can have on fostering ocean literacy and environmental literacy more broadly.

D. Edelist · D. Angel
University of Haifa, Haifa, Israel

Table 7.1 Ten dimensions of ocean literacy

Dimension of ocean literacy	*Description*
Knowledge	Knowledge has multiple aspects. In the first instance, knowledge is what a person knows about an ocean related topic and the links between topics. Knowledge also refers to the knowledge a person has about ocean decision making, opportunities to participate and engage in ocean decisions and behaviours and where/how to get information about ocean issues
Awareness	Awareness is the basic knowledge and understanding that a situation, problem or concept exists. Awareness should also include knowledge and understanding of the solutions and behaviours that may exist to address these problems in order to foster ownership and empower society to take action
Attitude	Attitude is related to a level of agreement with or concern for a particular position. Attitude should also include consideration of perceptions, values, and views towards an ocean issue, and how these can lead to policy and societal change
Behaviour	Behaviour relates to decisions, choices, actions, and habits with respect to ocean related issues at a range of scales, including from individual, sector and policy actors and institutions with a view to bringing about whole system change
Activism	Activism is the degree to which a person engages in a wide range of activities, which can constitute activism, such as campaigning (for example through social media, attending public rallies or writing to elected officials) to bring about changes in policy, attitudes, behaviour, etc. Understanding this dimension must also take account of who gets to participate in activism and what the barriers might be
Communication	Communication in the context of ocean literacy must be considered from multiple perspectives. (1) Communication is the extent to which a person communicates with others, such as family and peer groups, on ocean related topics. (2) Communication should also consider how/where people get their information about ocean issues from—What methods of communication are most effective? (3) At an organisational level, communication needs to consider how institutions and organisations are communicating to different audiences about ocean issues
Emotional connections	Emotional connections are about how a person feels and emotionally responds when they think about, are near/within, or consider issues relating to the ocean, coasts, and seas. Emotions can be positive, negative, or neutral and are all valid responses and will all contribute to behaviour change

(continued)

Table 7.1 (continued)

Dimension of ocean literacy	*Description*
Access and experience	Access and experience relate to a person's real or artificial (through Virtual Reality, for example) experiences and engagement with the ocean, and the various ways in which they can access these experiences. Barriers to ocean access and experiences should also be considered within this dimension
Adaptive capacity	Adaptive capacity relates to a person's capacity to adapt and respond to changing conditions relating to their ocean (e.g. relating to climate change, change in ocean economies, or changing ecosystem structure or function)
Trust and transparency	Trust and transparency relate to the level of trust a person places in sources of ocean information and knowledge, and their perception of how transparent information and associated platforms and processes are

Source McKinley et al. (2023, p. 3)

7.2 Connections Between Citizen Science and the Ten Dimensions of Ocean Literacy

7.2.1 Knowledge and Awareness

Knowledge and awareness are considered together in this chapter. Although knowledge and awareness are defined as two distinct dimensions of ocean literacy, their definitions overlap, and the concepts are often considered together or interchangeably in citizen science literature.

The potential for citizen science to impact participants' knowledge and awareness about a topic has been recognised for almost as long as the field of citizen science has existed itself (Bonney et al., 2009). A review of 35 articles reporting impacts of citizen science projects found that 18 articles (51%) considered participants' knowledge and, of these, 16 reported positive outcomes in participant knowledge (Stepenuk & Green, 2015). Similarly, a review of four categories of citizen science projects found limited but positive evidence that participants gain both knowledge and awareness when taking part in citizen science projects (Bonney et al., 2016). A number of marine plastic citizen science projects have recently monitored impacts on participants' knowledge and awareness all using the same survey adapted from Hartley et al. (2015). Two of the studies found increases in participants' knowledge about marine litter including more specific knowledge about the sources of marine litter e.g. a recognition that litter comes from businesses and fishing activities, and discharge from rivers, not just people dropping litter on beaches (Locritani et al., 2019; Severin et al., 2023). However, the other two studies found no significant change in participants' knowledge (Oturai et al., 2022; Wichmann et al., 2022). Furthermore, all four studies found limited or negligible differences in participants' awareness of marine litter. The authors suggest that these results may be due to participants' very high baseline knowledge and awareness of marine litter as any impact will inevitably be dependent on participants' level of knowledge prior to taking part. This is particularly plausible given the high societal awareness around marine plastic pollution (Baechler et al., 2024). Nevertheless, these examples complicate the notion that citizen science projects inevitably lead to an increase in participants' knowledge and awareness. Difficulties of generalising impacts across projects also arise due to the varying aims of citizen science projects. For example, there can be trade-offs between the education outcomes of a project and other objectives such as scientific outcomes and engagement of participants, each of which can be constrained by available time, space, staff and associated costs (Roche et al., 2020). Understanding of the impact of citizen science on participants' knowledge and awareness might also be limited by the assessment methods deployed and could be strengthened by adopting more comprehensive measures of knowledge gain (Somerwill & Wehn, 2022). For example, the three-dimensional theory of environmental knowledge, which considers system knowledge, action-related knowledge, and effectiveness knowledge, reflects the multiple elements in the ocean literacy knowledge dimension (Braun & Dierkes, 2019). Overall, there is a widely held assumption that citizen science positively impacts participants' knowledge and awareness and some evidence that this is true in specific contexts. However, the current evidence is not conclusive, and citizen science also appears to have limited impact on knowledge and awareness in other contexts. Furthermore, the timeframe of impact assessment is often shortly after the citizen science experience, potentially failing to assess whether changes in participants' knowledge and awareness are sustained beyond the end of the project (Wehn et al., 2021). Further evidence, utilising more comprehensive assessment methods, could provide insight into which citizen science approaches, contexts, and marine topics have the greatest potential for long-term impact on participants' knowledge and awareness.

7.2.2 *Attitudes*

Citizen science projects are assumed to foster positive attitudes toward science by increasing participants' knowledge and experience of research (Bruckermann et al., 2021). Citizen science has also been connected with the development of pro-environmental attitudes and increased engagement with conservation (Toomey & Domroese, 2013). In theory, citizen science could affect participants' attitudes on an even wider range of topics such as attitudes to specific organisations involved in the project or particular environmental management approaches. However, studies examining the effects of citizen science projects on attitudes are relatively uncommon within citizen science research and often present mixed results. A literature review of the impacts of citizen science studies on participants found only 15 out of 1240 citizen science articles reviewed (1%) considered changes in attitudes towards science (Finger et al., 2023). Of these, only one paper reviewed found a positive effect of citizen science on participants' attitudes, with the other studies finding no significant effect. This review only considered attitudes towards science but suggests that attitudes are difficult to change especially with only one-off or short-term participation in a project. In a recent marine example, participants in an Australian turtle mapping project, TurtleSAT, were surveyed; the results suggest a positive impact on participant's attitudes towards turtles (Santori et al., 2021). However, the conclusions were based on limited, self-reported data with questions such as "Have your attitudes towards turtles changed since you started participating in TurtleSAT?". As with knowledge and awareness, the conclusions drawn from these studies are often weakened by limitations in the methods deployed. Again, the review of impact-assessment methods by Somerwill and Wehn (2022) highlights the potential to adopt methods from the field of environmental psychology which has more developed approaches to measure environmental attitudes such as the Environmental Attitudes Inventory (Milfont & Duckitt, 2010). A high pre-existing level of positive attitudes among participants might also limit the effect of citizen science observed in these studies. As most citizen science projects recruit volunteers, the majority of participants self-select to take part in the project and might therefore be more likely to already have attitudes aligned with the aims of the project. This lack of diversity in the recruitment of participants could limit the impacts of citizen science (Pateman & West, 2023). Another pathway to increasing the impact of citizen science on attitudes could therefore be to focus recruitment on individuals who previously had negative or less positive attitudes towards the ocean and environment more broadly.

7.2.3 *Behaviour*

Participation in marine citizen science projects can be seen as a form of pro-environmental behaviour and, in this light, recruiting participants to a citizen science project can be seen as behaviour change and a route to fostering ocean literacy (Wyles & Ghilardi-Lopes, 2023). Beyond this intrinsic link, citizen science has also been proposed to contribute to further behaviour change outside of just participation in a project (van Noordwijk et al., 2021). Existing literature demonstrates some correlation between citizen science and broader behavioural change, although results have been mixed and the evidence so far is not conclusive (Phillips et al., 2018). For example, pre- and post-participation surveys with volunteers in a marine litter citizen science project showed that participants had higher intention to engage with pro-environmental behaviour, although this intention had decreased even just one week after taking part and the impact on behaviour from citizen science was not significantly different to other beach-based activities such as rock-pooling or coastal walking (Wyles et al., 2017). As with other dimensions of ocean literacy, there is overall a lack of evidence demonstrating the link between citizen science and behaviour change, including whether any change in behaviour is sustained in the long-term beyond participation in the project. For example, in a review of marine litter citizen science studies, fewer than 10% of the publications reported impacts on the participants such as behaviour change (Kawabe et al., 2022). This gap could be addressed with more frequent and comprehensive assessment of impacts on behaviour (including follow-up assessment of participants) and the adoption of more robust methodologies to evaluate behavioural outcomes (Somerwill & Wehn, 2022).

7.2.4 *Activism*

The role citizen science may play in activism and advocacy—whereby citizens intervene on behalf of a socio- or enviro-political goal—is a key consideration in its interactions with ocean literacy. Political advocacy, one of six key pathways through which citizen science is thought to be able to create positive environmental change (van Noordwijk et al., 2021), sees individuals and communities involved in citizen science inspired to publicly support causes and advocate for policy change. Such pressure can push issues up the political agenda and can lead to more rapid change than through presentation of new scientific evidence alone (Van Brussel & Huyse, 2019).

In a marine context, citizen science projects focused on plastic litter in the oceans provide a clear case of overlap between citizen science and activism. Beach cleaning, a common activity within litter-monitoring projects, can itself be considered a form of focused activism in a local space (Power, 2022). However, many marine litter citizen science projects attempt to go beyond this and also aim to increase participants' motivation to take action and push decision making (Kawabe et al., 2022). The connection between citizen science and activism is clear in cases like marine litter where many citizen science projects are organised as campaigns with strong emphasis on raising awareness, changing behaviour and increasing political will to tackle the issues of plastic pollution. It should be noted that citizen science projects are not always connected with activism and indeed it might not always be desirable as a politicised, activist approach may be in tension with other aims of the project and may alienate some participants (Roche et al., 2020).

7.2.5 *Communication*

As a dimension of ocean literacy, communication refers both to how individuals communicate with others and how they are communicated to by organisations. Citizen science projects are often collaborations between communities of individuals and organisations and, as such, they could improve both aspects of communication. Previous studies have highlighted the importance of citizen science projects in providing tools to facilitate communication between participants in order to support community building (Jennett et al., 2016). Social media can often play an important role in this process. For example, a study of participants in the marine citizen science project, Redmap, highlighted the importance of Facebook which was used by over 75% of participants to share and comment on each other's images of marine species (Martin et al., 2016).

Beyond peer-to-peer communication, citizen science often also enhances the way organisations communicate with the public and can help address the lack of communication between science, policy, and society

(Hecker et al., 2018). Citizen science is closely connected to the concept of open science with shared priorities around increasing the accessibility of research results and processes (Vohland & Göbel, 2017). For example, citizen science can improve science communication by foregrounding communication within the research process and sharing findings and results throughout the project rather than only at fixed points in the process, often only once the research is finished (Wagenknecht et al., 2021). This is also reflected in the ECSA 10 principles of citizen science which encourage not just the open sharing of data but also direct feedback and communication from the project to all its participants (Robinson et al., 2018).

7.2.6 Emotional Connections

Whilst enabling communities to take part in scientific research, citizen science projects often promote environmental engagement. One of the central elements of this engagement is related to emotional experiences (Reeve, 2018), which are a recognised source of motivation for continuous participation. Interestingly, both positive and negative emotions can play an important role in engaging citizen science participants. A survey of individuals attending coral reef citizen science events has shown that experiencing negative emotions about environmental problems is one of the key elements of the experience that contributes to greater levels of environmental engagement and support for marine conservation (Dean et al., 2018). In contrast, in the same survey, Dean et al. (2018) found that positive emotions, such as excitement, are associated with outcomes in individuals with low environmental identity (defined as the extent to which an individual sees themselves as an "environmental person"), or infrequent marine visitors. This suggests that positive emotions can be important for the initial stages of engagement of individuals with the environment, up to the level of transition to a higher degree of environmental identity.

As well as utilising emotion as the basis of engagement, citizen science projects can help to strengthen and broaden the emotional attachment participants have with the place they study. Empirical evidence for these people-place bonds was collected in a survey of participants in the citizen science project COASST (Coastal Observation and Seabird Survey Team) with results highlighting the importance that affinity for science can have in developing these emotional connections (Haywood et al., 2024).

Contribution to citizen science projects can also build a strong emotional sense of pride due to effort, time, and energy dedicated by participants (Greving et al., 2023).

7.2.7 Access and Experience

Participation in citizen science projects can be seen as a form of experience in itself, with many marine citizen science projects empowering participants to actively engage with the marine environment (Schuttler et al., 2018). In this sense, by granting participants a new experience, most citizen science projects address this dimension of ocean literacy to some extent. However, available data on diversity of participants in citizen science suggests accessibility is an issue in many projects with already marginalised societal groups less likely to participate (Pateman et al., 2021). In recognition of this, several citizen science projects and networks are actively attempting to increase accessibility establishing working groups and new methods of engagement (Paleco et al., 2021). However, in practice, there are still often issues of accessibility in citizen science including limited access to required technology, difficulties with the data collection procedure and platform, and physical limitations which make certain monitoring sites inaccessible (Lynch & Miller, 2023; Pocock et al., 2019). Ultimately, these issues of accessibility limit the impacts of citizen science, particularly with regards to which participants can benefit from the experience (Pateman & West, 2023). There is therefore a need to continue to build on work which explicitly considers accessibility in citizen science and explore solutions which give access to groups who are currently excluded.

7.2.8 Adaptive Capacity

Citizen science is put forward as a method for directly fostering adaptive capacity, extending science to help communities learn about, and adapt to, environmental hazards and change in the places they live (Bremer et al., 2019). A review of 44 studies focusing on the interface between citizen science and climate governance highlighted how citizen science can facilitate participants' adaptation—counteracting mental health distress and providing tools and skills to eliminate or reduce climate impacts and health-related risks (Freschi et al., 2024). In many cases, citizen science projects focus on monitoring a particular environmental hazard and any increases in participants' adaptive capacity are an added benefit. However, a community-led initiative related to tsunami preparedness in New Zealand provides a case in which the topic of the citizen science project itself was building adaptive capacity (Doyle et al., 2020). The project included a citizen-initiated survey whose results demonstrated that local residents lacked understanding on how best to respond to a tsunami. The survey results were also used to plan a local evacuation exercise including an evaluation of how participants now chose to respond. Whilst this case is perhaps unique in being so explicitly focused on adaptive capacity, it is nonetheless clear that citizen science can be used as a tool to foster resilience within communities. Beyond just individual adaptive capacity, citizen science can also contribute to collective adaptive capacity by providing data to inform decision making (Manteaw et al., 2022). In recognition of this, citizen science (as an indicator of more inclusive decision making) has been included as a criterion in a tool for evaluating regional adaptive capacity (Hirschfeld et al., 2020).

7.2.9 *Trust and Transparency*

Trust and transparency in relation to sources of information and knowledge, and the platforms and processes that disseminate it, is a key dimension when considering ocean literacy. Through its innate design of enabling the public to participate in research, citizen science methodologies have the potential to increase this trust. Previous citizen science activities have shown that participation of citizens in the research process can increase public trust in the results derived, especially when compared to non-participatory "traditional" research (Bedessem et al., 2021). Participating in citizen science can show a positive correlation with trust in researchers' honesty, trust in science as providing data for public policies, and trust in technical solutions to environmental problems (Bedessem et al., 2023).

Whilst the link between public trust in research and participating in the scientific process has been long established, there are different ways of achieving this through citizen science methodologies. Considering coastal and marine research specifically, the importance of high-quality, scientist-citizen communication has been reported, with different communication techniques and strategies required to engage diverse marine user groups (Hind-Ozan et al., 2017). The importance of communication when developing trust in research is not limited to scientist-citizen interactions, however. The marine space has a range of users and stakeholders, and citizen science can help foster common ground between them. Citizen science programmes can provide a site of "common ground" for users of the marine environment to connect and better understand each other. This in turn can improve not only trust in the research, but also regarding the needs and concerns of different user groups (Kelly et al., 2019). However, in such cases, careful attention has to be given to power-dynamics between different groups as assumptions of mutual trust and symmetrical power-relations can obscure how certain actors can have disproportionate influence over the design of collaborative, community science projects (McAteer & Flannery, 2022).

Elements relating to trust and transparency are a core part of ECSA's citizen science principles (Robinson et al., 2018), specifically regarding the open sharing of data and metadata, and the consideration of ethical issues regarding the research. By involving the public in scientific research, citizen science ubiquitously creates a communication channel between the scientist and the citizen, increasing awareness and knowledge regarding marine and ocean concerns, and trust in the methodologies used to study them. However, whilst the benefits of citizen science are thoroughly considered in ongoing research, negative impacts are seldom considered (Walker et al., 2021). More research is needed to consider any barriers to trust, and regarding the transparency of methods and data that can sometimes be more harmful than valued (Quinn, 2021).

7.3 Case Studies Illustrating How Citizen Science Projects View Environmental Literacy

7.3.1 ProBleu

ProBleu [https://probleu.school/] is a Horizon Europe EU project (2023–2026) which aims to boost ocean and water literacy by supporting the Network of European Blue Schools. Citizen science is central to the project methodology, envisaged as a tool to foster elements of ocean literacy across school communities, and the project plans to adapt and develop citizen science resources to facilitate the creation of blue school projects (Ceccaroni et al., 2023). Examples of the school projects supported can be found on the ProBleu website: https://probleu.school/funded-schools/. The project also recognises the importance of access and experience in achieving its intended societal impacts and, in particular, considers issues of accessibility around language, disability, and socio-economic barriers. As part of the project's evaluation process, ProBleu is developing an impact assessment methodology to assess learning outcomes from blue school projects, including changes in ocean and water literacy.

7.3.2 *Iliad*

Iliad [https://ocean-twin.eu/] is a Horizon 2020 EU project (2022–2025) which aims to demonstrate the technologies and methodologies required to develop Digital Twins of the Ocean (DTOs). The project is developing over 20 separate DTOs many of which include elements of citizen science and engagement with the goal of increasing public engagement in the DTO, increasing public participation in decision making, and fostering ocean literacy (Parkinson et al., 2024). One such example is the Jellyfish swarm forecast DTO which combines citizen science data from the Meduzot project (Edelist et al., 2022) with metocean parameters to develop an interactive forecasting tool for jellyfish swarms in Israel.

The Meduzot project addresses several aspects of ocean literacy. One aim of the project is to provide participants with knowledge to inform decision making such as how to avoid swarms of stinging jellyfish and the best treatment for stung bathers (Edelist et al., 2023). The project uses social media to facilitate communication to participants, and also between participants. The Meduzot Facebook group has over 15,000 members and citizen scientists often share photos, help each other with species identification, and ask each other questions about their experiences. To ensure trust and transparency, the project description is presented according to the CS Track guidelines (Golumbic & Oesterheld, 2023).

7.3.3 *FreshWater Watch*

Whilst the concept of ocean literacy focuses explicitly on the marine environment, the ten dimensions used to define ocean literacy can be applied to other environmental contexts and extended to consider environmental literacy more broadly, encompassing, for example, freshwater literacy. Under this umbrella, one can examine projects such as FreshWater Watch [https://www.freshwaterwatch.org/]; a global citizen science initiative run by Earthwatch Europe where communities are trained to monitor and protect their local rivers, lakes, and freshwater bodies. FreshWater

Watchers record basic visual observations of their local waterbody, the turbidity (i.e. how clear or cloudy the water is), and concentrations of two important macronutrients: nitrates and phosphates (Bishop et al., 2025).

As part of the FreshWater Watch initiative, over 2600 citizen scientists participated in a UK national WaterBlitz, testing the health of rivers, lakes, streams, ponds, and canals. Following the WaterBlitz, 528 participants responded to a questionnaire about their experiences (Bannatyne et al., 2024) evidencing dimensions of freshwater literacy. For example, 46% of participants had no prior involvement in the environmental sector before the WaterBlitz, suggesting the activity facilitated the access and experience dimension of freshwater literacy. The majority of participants either agreed (47%) or strongly agreed (13%) that their understanding of water quality issues and river health was improved through taking part in the WaterBlitz, evidencing improvements in knowledge and awareness. Furthermore, 22% reported a positive shift in their relationship with nature, indicative of the impact the WaterBlitz had on participant attitudes and perhaps even their emotional connection towards water.

7.4 Future Research Directions and Conclusions

This chapter has established the close connection between the aims of marine citizen science and ocean literacy and highlighted the potential for citizen science as a tool to bridge the gap between society and the ocean. The varied case studies presented demonstrate the wide diversity of contexts in which citizen science is applied, and the corresponding range of ways projects perceive their connection to environmental literacy. Links with literacy emerge, not just in education contexts such as in the ProBleu project where there is a deliberate focus on ocean literacy, but also as an impact in other citizen science projects where fostering environmental literacy among participants is not an explicit aim. To date, many marine citizen science projects have not explicitly considered ocean literacy within their project aims and objectives. A better understanding of the connection between citizen science and ocean literacy might therefore emerge if future citizen science projects and research more proactively engage with all the dimensions of ocean literacy.

A number of broad lessons learned emerged from this review. Firstly, the influence of citizen science on ocean literacy is not always obvious and evidence of the impacts of citizen science is often limited. Projects would benefit from tailored evaluation processes to measure the impacts of the project on each ocean literacy dimension. As is true of the ocean-literacy movement more broadly, improved impact assessment would help the marine citizen science community to identify best-practices in fostering literacy across all ten dimensions (Shellock et al., 2024). Citizen science is likely to be more impactful when deployed in conjunction with other engagement approaches as part of a wider ocean literacy campaign and impact will also be increased when projects are designed with specific dimensions of ocean literacy in mind. Future research can build on this review by working with citizen science projects to set objectives around fostering ocean literacy and providing practical frameworks to monitor their impacts in these areas.

Whilst this chapter has looked specifically at ocean literacy, the ten dimensions analysed are equally applicable to other domains (as demonstrated by the FreshWater Watch example). Citizen science can therefore be considered as a tool, not just for fostering ocean literacy, but also for environmental literacy more broadly.

References

Baechler, B. R., De Frond, H., Dropkin, L., Leonard, G. H., Proano, L., & Mallos, N. J. (2024). Public awareness and perceptions of ocean plastic pollution and support for solutions in the United States. *Frontiers in Marine Science, 10*, 1323477. https://doi.org/10.3389/fmars.2023.1323477

Bannatyne, L. et al. (2024). *The great UK WaterBlitz*. Earthwatch Europe. https://doi.org/10.5281/zenodo.12605436. Accessed 30 June 2024.

Bedessem, B., Dozières, A., Prévot, A. C., & Julliard, R. (2023). Science knowledge and trust in science in biodiversity citizen science projects. *JCOM: Journal of Science Communication, 22*(1), A05. https://doi.org/10.22323/2.22010205

Bedessem, B., Gawrońska-Nowak, B., & Lis, P. (2021). Can citizen science increase trust in research? A case study of delineating Polish metropolitan areas. *Journal of Contemporary European Research, 17*(2). https://doi.org/10.30950/jcer.v17i2.1185

Bishop, I., Boldrini, A., Clymans, W., Hall, C., Moorhouse, H., Parkinson, S., Scott-Somme, K., Thornhill, I., & Loiselle, S. (2025). Freshwater watch: Investigating the health of freshwater ecosystems, from the bottom up. *Citizen Science: Theory and Practice, 10*(1). https://doi.org/10.5334/cstp.754

Bonney, R., Cooper, C. B., Dickinson, J., Kelling, S., Phillips, T., Rosenberg, K. V., & Shirk, J. (2009). Citizen science: A developing tool for expanding science knowledge and scientific literacy. *BioScience, 59*(11), 977–984. https://doi.org/10.1525/bio.2009.59.11.9

Bonney, R., Phillips, T. B., Ballard, H. L., & Enck, J. W. (2016). Can citizen science enhance public understanding of science? *Public Understanding of Science, 25*(1), 2–16. https://doi.org/10.1177/0963662515607406

Braun, T., & Dierkes, P. (2019). Evaluating three dimensions of environmental knowledge and their impact on behaviour. *Research in Science Education, 49*(5), 1347–1365. https://doi.org/10.1007/s11165-017-9658-7

Bremer, S., Haque, M. M., Aziz, S. B., & Kvamme, S. (2019). "My new routine": Assessing the impact of citizen science on climate adaptation in Bangladesh. *Environmental Science & Policy, 94*, 245–257. https://doi.org/10.1016/j.envsci.2018.12.29

Bruckermann, T., Greving, H., Schumann, A., Stillfried, M., Börner, K., Kimmig, S. E., Hagen, R., Brandt, M., & Harms, U. (2021). To know about science is to love it? Unraveling cause–effect relationships between knowledge and attitudes toward science in citizen science on urban wildlife ecology. *Journal of Research in Science Teaching, 58*(8), 1179–1202. https://doi.org/10.1002/tea.21697

Cava, F., Schoedinger, S., Strang, C., & Tuddenham, P. (2005). *Science content and standards for ocean literacy: A report on ocean literacy*. http://www.cosee.net/files/coseeca/OLit04-05FinalReport.pdf. Accessed 29 August 2024.

Ceccaroni, L., Bowser, A., & Brenton, P. (2017). Civic education and citizen science: Definitions, categories, knowledge representation. In *Analyzing the role of citizen science in modern research* (pp. 1–23). IGI Global. https://doi.org/10.4018/978-1-5225-0962-2.ch001

Ceccaroni, L., Woods, S. M., Butkevičienė, E., Parkinson, S., Sprinks, J., Costa, P., Simis, S. G., Lessin, G., Liñán, S., Companys, B., & Bonfill, E. (2023). The role of citizen science in promoting ocean and water literacy in school communities: The ProBleu Methodology. *Sustainability, 15*(14), 11410. https://doi.org/10.3390/su151411410

Dean, A. J., Church, E. K., Loder, J., Fielding, K. S., & Wilson, K. A. (2018). How do marine and coastal citizen science experiences foster environmental engagement? *Journal of Environmental Management, 213*, 409–416. https://doi.org/10.1016/j.jenvman.2018.02.080

Doyle, E. E., Lambie, E., Orchiston, C., Becker, J. S., McLaren, L., Johnston, D., & Leonard, G. (2020). Citizen science as a catalyst for community resilience building: A two-phase tsunami case study. *Australasian Journal of Disaster and Trauma Studies, 24*(1), 23–50. http://trauma.massey.ac.nz/issues/2020-1/AJDTS_24_1_Doyle.pdf. Accessed 30 June 2024.

Edelist, D., Knutsen, Ø., Ellingsen, I., Majaneva, S., Aberle, N., Dror, H., & Angel, D. L. (2022). Tracking jellyfish swarm origins using a combined oceanographic-genetic-citizen science approach. *Frontiers in Marine Science, 9*. https://doi.org/10.3389/fmars.2022.869619.

Edelist, D., Angel, D. L., Barkan, N., Danino-Gozlan, C., Palanker, A., Barak, L., Robertson, E., & Bentur, Y. (2023). Jellyfish sting web survey: Clinical characteristics and management of *Rhopilema nomadica* envenomation in the Mediterranean Sea. *Regional Environmental Change, 23*(3), 114. https://doi.org/10.1007/s10113-023-02104-4

Finger, L., van den Bogaert, V., Schmidt, L., Fleischer, J., Stadtler, M., Sommer, K., & Wirth, J. (2023). The science of citizen science: A systematic literature review on educational and scientific outcomes. *Frontiers in Education, 8*. https://doi.org/10.3389/feduc.2023.1226529

Freschi, G., Menegatto, M., & Zamperini, A. (2024). Conceptualising the link between citizen science and climate governance: A systematic review. *Climate, 12*(5), 60. https://doi.org/10.3390/cli12050060

Garcia-Soto, C., Seys, J. J., Zielinski, O., Busch, J. A., Luna, S. I., Baez, J. C., Domegan, C., Dubsky, K., Kotynska-Zielinska, I., Loubat, P., & Malfatti, F. (2021). Marine citizen science: current state in Europe and new technological developments. *Frontiers in Marine Science, 8*. https://doi.org/10.3389/fmars.2021.621472

Glithero, L. D., Bridge, N., Hart, N., Mann-Lang, J., McPhie, R., Paul, K., Peebler, A., Wiener, C., Yen, C., Kelly, R., & McRuer, J. (2024). Ocean *Decade vision 2030 white papers: Challenge 10: Restoring society's relationship with the ocean*. UNESCO-IOC. https://doi.org/10.25607/ekwn-wh61. Accessed 29 August 2024.

Golumbic, Y. N., & Oesterheld, M. (2023). From goals to engagement—Evaluating citizen science project descriptions as science communication texts. *Frontiers in Environmental Science, 11*. https://doi.org/10.3389/fenvs.2023.1228480.

Greving, H., Bruckermann, T., Schumann, A., Stillfried, M., Börner, K., Hagen, R., Kimmig, S. E., Brandt, M., & Kimmerle, J. (2023). Attitudes toward engagement in citizen science increase self-related, ecology-related, and motivation-related outcomes in an urban wildlife project. *BioScience, 73*(3), 206–219. https://doi.org/10.1093/biosci/biad003

Hartley, B. L., Thompson, R. C., & Pahl, S. (2015). Marine litter education boosts children's understanding and self-reported actions. *Marine Pollution Bulletin, 90*(1), 209–217. https://doi.org/10.1016/j.marpolbul.2014.10.049

Haywood, B. K., Parrish, J. K., Jones, T., & Inman, S. (2024). Shaping people-place bonds in citizen science: A framework for analysis. *Ecology and Society, 29*(1). https://doi.org/10.5751/ES-14754-290111

Hecker, S., Haklay, M., Bowser, A., Makuch, Z., Vogel, J., & Bonn, A. (2018). Stories can change the world—Citizen science communication in practice. In

S. Hecker et al. (Eds.), *Citizen science* (pp. 445–462). UCL Press (Innovation in Open Science, Society and Policy). https://doi.org/10.14324/111.9781787352339

Hind-Ozan, E. J., Pecl, G. T., & Ward-Paige, C. A. (2017). Communication and trust-building with the broader public through coastal and marine citizen science. In *Citizen science for coastal and marine conservation*. Routledge.

Hirschfeld, D., Hill, K. E., & Riordan, B. (2020). The regional fingerprint: A new tool to evaluate adaptive capacity. *Environmental Science & Policy, 112*, 36–46. https://doi.org/10.1016/j.envsci.2020.05.019

Jennett, C., Kloetzer, L., Schneider, D., Iacovides, I., Cox, A.L., Gold, M., Fuchs, B., Eveleigh, A., Mathieu, K., Ajani, Z., & Talsi, Y. (2016). Motivations, learning and creativity in online citizen science. *Journal of Science Communication, 15*(3). https://doi.org/10.22323/2.15030205

Kawabe, L. A., Ghilardi-Lopes, N. P., Turra, A., & Wyles, K. J. (2022). Citizen science in marine litter research: A review. *Marine Pollution Bulletin, 182*, 114011. https://doi.org/10.1016/j.marpolbul.2022.114011

Kelly, R., Fleming, A., & Pecl, G. T. (2019). Citizen science and social licence: Improving perceptions and connecting marine user groups. *Ocean & Coastal Management, 178*, 104855. https://doi.org/10.1016/j.ocecoaman.2019.104855

Locritani, M., Merlino, S., & Abbate, M. (2019). Assessing the citizen science approach as tool to increase awareness on the marine litter problem. *Marine Pollution Bulletin, 140*, 320–329. https://doi.org/10.1016/j.marpolbul.2019.01.023

Lynch, L., & Miller, J. (2023). Accessibility of participation in a pollinator-focused community science project. *Citizen Science: Theory and Practice, 8*(1). https://doi.org/10.5334/cstp.557

MacNeil, S., Hoover, C., Ostertag, J., Yumagulova, L., & Glithero, L. D. (2021). Coming to terms with ocean literacy. *Canadian Journal of Environmental Education (CJEE), 24*(1), 233–252. https://cjee.lakeheadu.ca/article/view/1777. Accessed 5 June 2024.

Manteaw, B. O., Amoah, A. B., Ayittah, B., & Enu, K. B. (2022). Climate-Informed decision-making in data-poor environments: Managing climate risk through citizen science networks. *Frontiers in Climate, 4*. https://doi.org/10.3389/fclim.2022.835768

Martin, V. Y., Christidis, L., Lloyd, D. J., & Pecl, G. (2016). Understanding drivers, barriers and information sources for public participation in marine citizen science. *Journal of Science Communication, 15*(2), A02. https://doi.org/10.22323/2.15020202.

McAteer, B., & Flannery, W. (2022). Power, knowledge and the transformative potential of marine community science. *Ocean & Coastal Management, 218*, 06036. https://doi.org/10.1016/j.ocecoaman.2022.106036

McKinley, E., Burdon, D., & Shellock, R. J. (2023). The evolution of ocean literacy: A new framework for the United Nations Ocean Decade and beyond. *Marine Pollution Bulletin, 186*, 114467. https://doi.org/10.1016/j.marpolbul.2022.114467

Milfont, T. L., & Duckitt, J. (2010). The environmental attitudes inventory: A valid and reliable measure to assess the structure of environmental attitudes. *Journal of Environmental Psychology, 30*(1), 80–94. https://doi.org/10.1016/j.jenvp.2009.09.001

Oturai, N. G., Pahl, S., & Syberg, K. (2022). How can we test plastic pollution perceptions and behavior? A feasibility study with Danish children participating in "the Mass Experiment." *Science of the Total Environment, 806*, 150914. https://doi.org/10.1016/j.scitotenv.2021.150914

Paleco, C., García Peter, S., Salas Seoane, N., Kaufmann, J., Argyri, P., Vohland, K., Land-Zandstra, A., Ceccaroni, L., Lemmens, R., Perelló, J., & Ponti, M. (2021). Inclusiveness and diversity in citizen science. In K. Vohland et al. (Eds.), *The science of citizen science* (pp. 261–281). Springer International Publishing. https://doi.org/10.1007/978-3-030-58278-4_14

Pateman, R., Dyke, A., & West, S. (2021). The diversity of participants in environmental citizen science. *Citizen Science: Theory and Practice, 6*(1). https://doi.org/10.5334/cstp.369

Pateman, R. M., & West, S. E. (2023). Citizen science: Pathways to impact and why participant diversity matters. *Citizen Science: Theory and Practice, 8*(1). https://doi.org/10.5334/cstp.569

Phillips, T., Porticella, N., Constas, M., & Bonney, R. (2018). A framework for articulating and measuring individual learning outcomes from participation in citizen science. *Citizen Science: Theory and Practice, 3*(2). https://doi.org/10.5334/cstp.126

Pocock, M. J., Roy, H. E., August, T., Kuria, A., Barasa, F., Bett, J., Githiru, M., Kairo, J., Kimani, J., Kinuthia, W., & Kissui, B. (2019). Developing the global potential of citizen science: Assessing opportunities that benefit people, society and the environment in East Africa. *Journal of Applied Ecology, 56*(2), 274–281. https://doi.org/10.1111/1365-2664.13279

Power, S. (2022). Enjoying your beach and cleaning it too: A grounded theory ethnography of enviro-leisure activism. *Journal of Sustainable Tourism, 30*(6), 1438–1457. https://doi.org/10.1080/09669582.2021.1953037

Quinn, A. (2021). Transparency and secrecy in citizen science: Lessons from herping. *Studies in History and Philosophy of Science Part A, 85*, 208–217. https://doi.org/10.1016/j.shpsa.2020.10.010

Parkinson, S., Ceccaroni, L., Edelist, D., Robertson, E., Horincar, R., Laudy, C., Ganchev, T., Markova, V., Pearlman, J., Simpson, P., Venus, V., Muchada, P., Kazanjian, G., Bye, B. L., Oliveira, M., Paredes, H., Sprinks, J., Witter, A., Cruz, B., ... Woods, S. M. (2024). The Iliad digital twins of the ocean:

opportunities for citizen science. *ARPHA Proceedings, 6*, 61–65. https://doi.org/10.3897/ap.e126643

Reeve, J. (2018). *Understanding motivation and emotion.* Wiley Custom.

Robinson, L. D., Cawthray, J. L., West, S. E., Bonn, A., & Ansine, J. (2018). Ten principles of citizen science. In A. Bonn et al. (Eds.), Citizen science (pp. 27–40). UCL Press (Innovation in Open Science, Society and Policy). https://doi.org/10.14324/111.9781787352339

Roche, J., Bell, L., Galvão, C., Golumbic, Y.N., Kloetzer, L., Knoben, N., Laakso, M., Lorke, J., Mannion, G., Massetti, L., & Mauchline, A. (2020). Citizen science, education, and learning: Challenges and opportunities. *Frontiers in Sociology, 5.* https://doi.org/10.3389/fsoc.2020.613814

Sandahl, A., & Tøttrup, A. P. (2020). Marine citizen science: recent developments and future recommendations. *Citizen Science: Theory and Practice, 5*(1). https://doi.org/10.5334/cstp.270

Santori, C., Keith, R. J., Whittington, C. M., Thompson, M. B., Van Dyke, J. U., & Spencer, R. J. (2021). Changes in participant behaviour and attitudes are associated with knowledge and skills gained by using a turtle conservation citizen science app. *People and Nature, 3*(1), 66–76. https://doi.org/10.1002/pan3.10184

Schuttler, S. G., Sorensen, A. E., Jordan, R. C., Cooper, C., & Shwartz, A. (2018). Bridging the nature gap: Can citizen science reverse the extinction of experience? *Frontiers in Ecology and the Environment, 16*(7), 405–411. https://doi.org/10.1002/fee.1826

Severin, M.I., Akpetou, L.K., Annasawmy, P., Asuquo, F.E., Beckman, F., Benomar, M., Jaya-Ram, A., Malouli, M., Mees, J., Monteiro, I., & Ndwiga, J. (2023). Impact of the citizen science project COLLECT on ocean literacy and well-being within a north/west African and south-east Asian context. *Frontiers in Psychology, 14.* https://doi.org/10.3389/fpsyg.2023.1130596

Shellock, R. J., Fullbrook, L., McKinley, E., Cvitanovic, C., Kelly, R., & Martin, V. (2024). The nature and use of ocean literacy in achieving sustainable ocean futures: A systematic map. *Ocean & Coastal Management, 257*, 107325. https://doi.org/10.1016/j.ocecoaman.2024.107325

Somerwill, L., & Wehn, U. (2022). How to measure the impact of citizen science on environmental attitudes, behaviour and knowledge? A review of state-of-the-art approaches. *Environmental Sciences Europe, 34*(1), 18. https://doi.org/10.1186/s12302-022-00596-1

Stepenuck, K. F., & Green, L. T. (2015). Individual- and community-level impacts of volunteer environmental monitoring: A synthesis of peer-reviewed literature. *Ecology and Society, 20*(3). https://doi.org/10.5751/ES-07329-200319

Toomey, A. H., & Domroese, M. C. (2013). Can citizen science lead to positive conservation attitudes and behaviors? *Human Ecology Review, 20*(1), 50–62. https://www.jstor.org/stable/24707571

Van Brussel, S., & Huyse, H. (2019). Citizen science on speed? Realising the triple objective of scientific rigour, policy influence and deep citizen engagement in a large-scale citizen science project on ambient air quality in Antwerp. *Journal of Environmental Planning and Management, 62*(3), 534–551. https://doi.org/10.1080/09640568.2018.1428183

van Noordwijk, T., Bishop, I., Staunton-Lamb, S., Oldfield, A., Loiselle, S., Geoghegan, H., & Ceccaroni, L. (2021). Creating positive environmental impact through citizen science. In K. Vohland et al. (Eds.), *The science of citizen science* (pp. 373–395). Springer International Publishing. https://doi.org/10.1007/978-3-030-58278-4_19

Vohland, K., & Göbel, C. (2017). Open Science und Citizen Science als symbiotische Beziehung? Eine Gegenüberstellung von Konzepten. *TATuP—Journal for Technology Assessment in Theory and Practice, 26*(1–2), 18–24. https://doi.org/10.14512/tatup.26.1-2.18

Wagenknecht, K., Woods, T., Nold, C., Rüfenacht, S., Voight-Heucke, S., Caplan, A., Hecker, S., & Vohland, K. (2021). A question of dialogue? Reflections on how citizen science can enhance communication between science and society. *Journal of Science Communication, 20*(3), p. A13. https://doi.org/10.22323/2.20030213.

Walker, D. W., Smigaj, M., & Tani, M. (2021). The benefits and negative impacts of citizen science applications to water as experienced by participants and communities. *Wires Water, 8*(1), e1488. https://doi.org/10.1002/wat2.1488

Wehn, U., Gharesifard, M., Ceccaroni, L., Joyce, H., Ajates, R., Woods, S., Bilbao, A., Parkinson, S., Gold, M., & Wheatland, J. (2021). Impact assessment of citizen science: state of the art and guiding principles for a consolidated approach. *Sustainability Science, 16*(5), 1683–1699. https://doi.org/10.1007/s11625-021-00959-2.

Wehn, U., Bilbao Erezkano, A., Somerwill, L., Linders, T., Maso, J., Parkinson, S., Semasingha, C., & Woods, S. (2025). Past and present marine citizen science around the globe: A cumulative inventory of initiatives and data produced. Ambio. https://doi.org/10.1007/s13280-024-02119-z

Wichmann, C. S., Fischer, D., Geiger, S. M., Honorato-Zimmer, D., Knickmeier, K., Kruse, K., Sundermann, A., & Thiel, M. (2022). Promoting pro-environmental behavior through citizen science? *A Case Study with Chilean Schoolchildren on Marine Plastic Pollution', Marine Policy, 141*, 105035. https://doi.org/10.1016/j.marpol.2022.105035

Wyles, K. J., Pahl, S., Holland, M., & Thompson, R. C. (2017). Can beach cleans do more than clean-up litter? Comparing beach cleans to other coastal activities. *Environment and Behavior, 49*(5), 509–535. https://doi.org/10.1177/0013916516649412

Wyles, K. J., & Ghilardi-Lopes, N. P. (2023). Citizen science as a pro-environmental behaviour and a catalyst for further behaviour change. In *Handbook on pro-environmental behaviour change*. Edward Elgar Publishing. https://doi.org/10.4337/9781800882133.00031

CHAPTER 8

Connecting People and the Ocean Through Street Art and Citizen Science

Baiba Prūse, Marta Dieviņa, Ribards Bražinskis, Anna Stolyarova, Noel Pérez, Cika Schulz, Geertje Tijsma, Svea Rogge, Tanja Straka, Marien Helmus, Embla Ekström, Ignacio Gianelli, Katina Roumbedakis, Natali Lazzari, Ludovica Montecchio, Milena Arias Schreiber, Arne Kinds, Philipp Hummer, Estefanía Sánchez-Flores, Rubén Vinueza-Chérrez, Sebastian Villasante, Eduardo Urias, and Sophia Kochalski

8.1 Background

> *"Open that window, I want to see the sea"*
> —Last words of Rosalía de Castro, Galician poet and novelist
> (María, 1991, p. 320)

Inspired by the ocean and world-renowned artists, our author team engages in the intersection between oceans, seas, and street art. We embrace the perspective that "the arts can advance sustainability science

B. Prūse (✉) · G. Tijsma · M. Helmus · E. Urias
Vrije Universiteit Amsterdam Athena Institute, Amsterdam, The Netherlands
e-mail: bpruse@ucc.ie

C. Certomà (ed.), *Blue Kinships*,
https://doi.org/10.1007/978-3-031-78619-8_8

by embracing transdisciplinarity and expanding conventional epistemologies towards practical, embodied, and emotional domains" (Heras et al., 2021). Murals, a prevalent form of street art, are increasingly recognised

G. Tijsma
e-mail: g.tijsma@vu.nl

M. Helmus
e-mail: marien.a.helmus@gmail.com

E. Urias
e-mail: eduardo.munizpereiraurias@vu.nl

B. Prūse
University College Cork, Cork, Ireland

M. Dieviņa
Latvian Academy of Culture, Riga, Latvia
e-mail: marta.dievina@gmail.com

R. Bražinskis
Independent Researcher of Graffiti and Street Art, Riga, Latvia
e-mail: brazinskis.rihards@gmail.com

A. Stolyarova
Street Art Museum Amsterdam, Amsterdam, The Netherlands
e-mail: answers@streetartmuseumamsterdam.com

N. Pérez
Viladomar Public Art Platform, Rianxo, Spain
e-mail: hola@novenoel.com

C. Schulz
Freelance Artist and Educator, Utrecht, The Netherlands
e-mail: cika.schulz@gmail.com

S. Rogge
Technische Universität Berlin, Berlin, Germany
e-mail: svea.rogge@campus.tu-berlin.de

T. Straka
Freie Universität Berlin, Berlin, Germany
e-mail: tanja.straka@fu-berlin.de

M. Helmus
University of Amsterdam, Amsterdam, The Netherlands

as tools to foster community engagement and raise environmental literacy and awareness (e.g., Sanchez et al., 2020; Schneller & Irizarry, 2014). Additionally, in recent years, the dialogue about citizen science initiatives

E. Ekström
Swedish Institute for the Marine Environment, University of Gothenburg, Gothenburg, Sweden
e-mail: embla.ekstrom@gu.se

I. Gianelli · N. Lazzari · M. Arias Schreiber · A. Kinds · S. Villasante · S. Kochalski
EqualSea Lab-CRETUS, Universidade de Santiago de Compostela, Santiago de Compostela, Spain
e-mail: ignaciogianelli@gmail.com

N. Lazzari
e-mail: lazzari.natali@gmail.com

M. Arias Schreiber
e-mail: milena.schreiber@gu.se

A. Kinds
e-mail: arnekinds@gmail.com

S. Villasante
e-mail: sebastian.villasante@usc.es

S. Kochalski
e-mail: kochalski.fisheries@gmail.com

I. Gianelli
South American Institute for Resilience and Sustainability Studies, Bella Vista, Uruguay

K. Roumbedakis
Instituto de Investigaciones Marinas (IIM-CSIC), Ecología Marina Integrada, Vigo, Spain

Centro de Investigación y de Estudios Avanzados (Cinvestav-Mérida), Mérida, México

K. Roumbedakis
e-mail: katina.roumbedakis@gmail.com

N. Lazzari
College of Science and Engineering, James Cook University, Townsville, QLD, Australia

within the humanities and social science has intensified (Smaniotto, 2023; Tauginienė et al., 2020), including public engagement with the arts in research (DITOs, 2019). Emerging initiatives integrate artistic practices in citizen science projects (Eco-arts project, n.d.), apply art for explorations (ARTigo, n.d.), merge citizen scientists' contributions into artistic displays (focusTerra, 2020), or serve as dialogue starters (Bonhoure et al., 2019; Sea Walls, n.d.).

Our approach centres on murals as a means of community engagement (Picture 8.1). As artist Christina Huynh states: "Street art and muralism bring local residents and the overall community of a place together" (Huynh & Jürgens, 2024). Despite this growing interest, few studies, such as Sanchez et al. (2020), have explored how street art, particularly murals, can be leveraged to enhance community engagement, raise awareness, and promote ocean conservation through the integration of citizen science.

To address this gap, this chapter shares insights from the international citizen science initiative "From Sea to Street" (FStS), which integrates street art, citizen science, and ocean conservation. Co-developed by three research institutions in Spain, Latvia, and the Netherlands, FStS leverages the power of murals to evoke emotions, inspire, and reshape perspectives on ocean conservation (Prūse et al., 2024). Through a series of

L. Montecchio
Ca' Foscari University of Venice, Venezia, Italy
e-mail: l.montecchio@campus.unimib.it

University of Milan-Bicocca, Milan, Italy

M. Arias Schreiber
School of Global studies, University of Gothenburg, Gothenburg, Sweden

P. Hummer
SPOTTERON Citizen Science App Platform, Vienna, Austria
e-mail: philipp@spotteron.net

E. Sánchez-Flores · R. Vinueza-Chérrez
Corporación Yemanyá—Agua y Conservación, Esmeraldas, Ecuador
e-mail: estefaniasf109@gmail.com

R. Vinueza-Chérrez
e-mail: rubencho_230@hotmail.com

Picture 8.1 A mural in Ladeira do chazo, Boiro, A Coruña, Spain by Novenoel (Spanish Artist). Copyright: with permission by the artist

reflections, the authors collectively explored three common themes that tied the experiences together, namely: (1) murals' contribution to ocean conservation and awareness, (2) citizen science—building a community; (3) lessons learned towards cultivating ocean identities.

8.2 Our Approach of Reflections

We invited a diverse group, including street artists, tour guides, students, staff of street art museums and citizen science apps, and project members and participants of FStS to share their viewpoints based on their own experiences within and beyond FStS. A conscious choice was made to seek additional input from authors with relevant experience in similar local projects who were not directly linked with FStS. We intended to put the experience of each contributor central stage and form a dialogue with the reader. The following sections present reflections based on three recurring elements and their intersections with each other: murals, citizen science, and ocean conservation. We follow the reasoning shared by Eisner (2003) where "[a]rtistic approaches to research are less concerned with the discovery of truth than with the creation of meaning".

8.3 Theme One: Murals and Its Contribution to Ocean Conservation and Awareness

8.3.1 Arne Kinds (Fisheries Researcher)

"Imagine an art practice that, instead of delighting merely the refined sensibilities of an elite few, has the power to engage, effortlessly and aesthetically, the masses through its manifest creativity, skill, originality, depth of meaning, and beauty". This is part of Nicholas Riggle's attempt at defining *street art* (2010), which he combines with elements of anonymity and ephemerality, as well as the impossibility for street art to be owned and appropriated.

The definition resonated with me. Owing to their size, bright colours, and engaging themes, the murals studied during this project are undeniable and in your face. And yet, we engage with them effortlessly, with varying levels of appreciation depending on our own reasons to be in the street, our mood, the weather. The artist leaves us to our senses and is not present to receive our praise or criticism. The painting just is.

But it is also much more than just that. Art has always been a vehicle for making scientific insights available to the wider population, including through music, film, and plastic arts. Regular exposure to ocean-themed art can contribute to ocean awareness, which is on the rise in societies all over the world (Worm et al., 2021). Many ocean-themed murals transcend mere aesthetics; often the artist depicts hot topics in marine science, ranging from overfishing to plastic pollution and global warming. As such, the artworks serve as a public outcry and manifesto advocating for a healthy and sustainable ocean, and the informed artist stands as a vital ally of marine science and conservation.

8.3.2 Novenoel (Street Artist)

As a mural painter with over 15 years of experience creating art in public spaces, I have found that paintings can have great communicative value and emotional impact on viewers. This was further confirmed during FStS conference at the Viladomar 2023 festival in Rianxo. Murals can generate questions and motivations for those who observe and interpret them. However, it is regularly necessary to provide an explanation or context for the work in order to fully understand its message. This can be done in-situ or via QR code. Despite this, murals have the ability to connect with audiences from diverse social backgrounds, who are often sensitive to the environment and its issues. The power of murals extends beyond their aesthetic appeal. They can serve as catalysts for debate, collective memory, and social action. This potential is particularly significant when it comes to transmitting informative and educational messages about environmental conservation. By engaging with the world of art, people from all walks of life can be inspired to take action and contribute to the preservation of our oceans and marine environment.

8.3.3 Katina Roumbedakis (Fisheries Researcher)

Arts in general—and, specifically, murals—have the ability to engage people and communities not only by capturing viewers' attention and curiosity due to their visual impact, often bringing beauty to urbanised areas, but also by creating emotional and cultural connection. Moreover, eventually, murals contain educational messages. In the context of the oceans, murals can be powerful tools for inspiring individuals to take action in conservation activities or generate awareness of the challenges the oceans face.

I have always been fascinated by mural arts (and artists) and can call myself a mural hunter for a long time. During the FStS project, I was able to explore two of my passions to help understand how murals and muralists represent the oceans. The project was a great opportunity to connect with a diverse group of researchers, muralists, and the general public.

During a mural tour at the Viladomar 2023 festival, we had the chance to reflect about the murals' message and have some "explained". Although arts are very much free interpretation, I really loved to hear the explanation of how the artists inspire themselves to produce the murals, all previous work and the mural painting itself, and the message from the artist's point of view. Although, in general, I was able to "get the message", seeing beyond the paint and learning about the artist's perspectives and life experiences made me change the way I see and interpret murals myself. Now, I am much more attentive to each detail, finding a new interpretation—and inspiration—at every time.

8.3.4 *Embla Ekström (Master's Graduate)*

I grew up in a small, inland town in Sweden where streets are accompanied by wooden or brick houses, and most of the aesthetics is housed by the facade of the buildings. Every other summer, visiting artists come to decorate the city in art, one of them painted a mural over one of the old factory walls, and my 15-year-old interest was caught. Suddenly, there was a large narwhal entangled in a fish line painted all over this wall that I regularly passed without giving a mere thought to the environmental issues hovering over the seas. It sparked an interest I had not yet been able to explore, by evoking emotions connected to nature. As it continues to present this opportunity for people that by routine or coincidence pass that wall, it invites inhabitants of this inland town to connect to the marine sphere, bridging the gap between nature perception and urbanised communities. The narwhal experience made a mark on the wall as well as in my mind. When I, years later, arrived in Galicia to assist in the project FStS I was about to finalise my master's in "Sea and society". Reflecting on what one singular ocean-related mural could give and seeing that the occurrence of those are rapidly increasing globally. Maybe the potential for murals to contribute to ocean literacy is solely limited to the occurrence of walls, which we have plenty of.

8.3.5 *Ignacio Gianelli (Sustainability Scientist)*

When I arrived in Galicia, I was struck by the richness and variety of murals on the streets, both in urban and rural areas. As a "mural hunter" and marine researcher, I soon realised that many of these murals contained elements related to the sea, its creatures or people. I asked myself, would it be possible to study marine biodiversity and sea-related traditions by analysing the content of murals? I would say yes. I believe that art in general, and murals in particular, have an incredible potential to raise awareness of any issue, including marine conservation, by portraying objects, entities, ways of life, and emotions attractively and inclusively. Recently, a cultural movement has emerged, transforming urban spaces into open-air museums and democratising art access on a large scale.

This movement has gained global recognition for its ability to bring about urban transformation through art, and is particularly appealing to the youth. At a time when our connection to nature is at a minimum, ocean-related murals open windows to the sea in the middle of cities or towns, reminding us (either consciously or unconsciously) of our intrinsic connection to nature. It may spark a conversation or a memory about the sea, or feed the curiosity to know more about a colourful sea creature embedded in the wall. Is it real? Does it live in our waters?

In FStS project, I had the opportunity to interview people and witness how a conversation about a mural could move a person to tears. I realised that some artists have integrated the sea into their art to such an extent that they do not notice the abundance of sea-related elements in Galician murals. For them, it was just their normality.

8.3.6 Sebastian Villasante (Interdisciplinary Scientist)

I work on sustainability transformations across multiple dimensions of human society needed to achieve sustainable futures on Earth and ocean. I am particularly interested in the use of imagination and creativity to move beyond familiar patterns to find new ways to understand our relation with the oceans. In particular, I found myself learning how shifts in mindsets and opening channels to include different perspectives and ways of knowing, being, and doing can help to deliberately activate initiatives that focus on the use of art as an inspiration to evoke transformative changes across a variety of dimensions of society. In the FStS project, I was able to collect, analyse, and reflect on how artists are able to use their imagination to create a mosaic of diverse expressions of how humans connect with the oceans in Galicia.

8.3.7 Tanja Straka (Researcher in Urban Ecology)

In my venture into ocean-themed murals, I have found myself captivated by the question of the team about the potential of murals to serve as a tool for ocean awareness-raising. As an urban ecologist, my interest is normally in understanding human-nature relationships (including human emotions) in terrestrial systems. However, in this project I had the chance to move beyond this. Reading the different comments from our survey participants, I was struck by the diverse range of reactions to the murals we presented. Their words conveyed a mix of awe, interest, contentment, and sometimes also amusement. Yet, among this range of emotions, I found myself reflecting on the original intentions of the artists. Did they aim to foster ocean awareness with their murals? What was their intention with the mural art? Undoubtedly, art has the power to evoke powerful emotions. It is rare to leave an exhibition without feelings of enlightenment, inspiration, or deep reflection. Reflecting on this, I now question whether murals alone suffice in raising awareness, or if their impact could be even enhanced through integration with education. Research supports the role of emotions in learning, indicating that they help in encoding information effectively and enhancing its retrieval process (Tyng et al., 2017). Consequently, I conclude that murals can serve as a gateway to an emotional connection with the ocean and if the intention is to promote conservation, incorporating informative signage alongside murals could further enhance their effectiveness as tools for communication.

8.4 Theme Two: Citizen Science—Building a Community and Dialogue

8.4.1 *Estefanía Sánchez Flores & Rubén Vinueza-Chérrez (Researchers/Coordinators of "Arte por el Agua")*

Citizen science is a cutting-edge tool in environmental education to promote pro-environmental behaviours. Its approach, where the community is part of the research, breaks the barrier between the teacher/researcher and the student, creating a meaningful and horizontal teaching and learning experience. It has used various methods that involve photography, digital applications, taxonomic guides, highlighting its interdisciplinary nature. Art is also an educational tool and a universal means of communication that complements citizen science. Through art, knowledge is acquired and environmental awareness and aesthetic sensitivity towards nature are developed. This, in turn, will channel more complex and multidimensional processes into people's subconscious, leading to the adoption of pro-environmental behaviours and sustainable practices. Projects where mural art has been the protagonist, such as "Arte por el Agua" and FStS are examples of this. "Arte por el Agua" developed in Ecuador, allowed for the rescue and exchange of knowledge and values about ecosystem services, marine biodiversity, and environmental issues. In addition, the present and future perspectives of the communities regarding their identity and territory were catalysed, democratising

Picture 8.2 A mural in the process of being completed in 2017 by residents of Playa Las Palmas in the province of Esmeraldas-Ecuador by Estefanía Sánchez Flores. Copyrights: with the permission of Estefanía Sánchez Flores

the participation of different people, with the mural becoming a space for exchange and connection. The interdisciplinary team that guided the project, composed of biologists, artists, cultural managers, and educators, achieved these objectives thanks to their experience and the strengths of each discipline. Finally, the creation of interactive murals, where passers-by had the opportunity to observe the artwork and interact with it through messages, expanded the artistic, educational, and information-gathering experience. These messages placed on the murals also helped understand how citizens perceive their natural environment and the challenges they face (Picture 8.2).

8.4.2 Baiba Prūse (Bee-Keepers Daughter/Researcher/Educator)

FStS offered an opportunity to experiment, to learn from each other without predefined answers. Isn't that what citizen science is also about? To learn, to share and above all—communicate (see Rüfenacht et al., 2021). Along this journey of FStS, we were able to embrace this learning journey and truly trust the process (during the project and beyond) where the role is not given but each and everyone is able to offer and therefore co-create the activities. Can a small initiative become a platform for like-minded people across the globe to implement hands-on activities? My answer: yes, it can.

8.4.3 Anna Stolyarova (Founder of Street Art Museum Amsterdam)

As the founder of the Street Art Museum Amsterdam, I have witnessed first-hand how citizen science projects serve as remarkable platforms where multiple disciplines converge, fostering collaborative interactions among various stakeholders. Our project, FStS, exemplifies this interdisciplinary synergy, integrating oceanography, environmental science, art, and education to address issues of sustainability and environmental awareness.

Through this initiative, we have created a vibrant dialogue between scientists, artists, educators and local communities. Street art tour guides and local community members contribute their cultural and contextual insights, ensuring that the scientific narratives embedded in the art are meaningful and relevant to the public. Their involvement bridges the gap between the scientific community and the general public, promoting broader environmental stewardship.

Educators leverage these artistic representations to create engaging, real-world learning experiences. By bringing students to view and interact with the street art, they transcend traditional classroom boundaries and immerse learners in the tangible impacts of environmental science. This experiential approach fosters a deeper connection to the subject matter and inspires a sense of responsibility towards our planet.

In summary, the FStS project illustrates how citizen science can act as a melting pot for diverse disciplines. It blends scientific rigour with artistic expression and community engagement, creating a holistic approach to environmental education and conservation. Through these multidisciplinary interactions, we not only enhance public understanding of marine ecosystems but also build a stronger, more connected community dedicated to protecting our blue planet.

8.4.4 Geertje Tijsma and Eduardo Urias (Researchers/Educators)

In 2023, we decided to add the element of street art and murals as part of the Interdisciplinary Community Service Learning course in order to facilitate the co-agenda setting process between students from different disciplines and residents from developing neighbourhoods in Amsterdam.

Engaging with art in collaboration with Street Art Museum Amsterdam (SAMA) has fostered students' awareness of societal issues related to urban transitions and has challenged their own views about art and the role that art plays in our society. Also, making street art together has facilitated knowledge sharing and conceptual disambiguation among students, as well as it has contributed positively to their team dynamics by creating a safe space to share and be creative. Furthermore, we found that the nature of the community partner, namely SAMA, who applies street art as a tool for dialogue between stakeholders, enabled and inspired the learnings of the students because they truly practised the combination of creativity and community building. We are grateful for this learning as it provided a new view on our educational practices, and we will strive for continued use of street art to facilitate the interactions between various (disciplinary) backgrounds.

8.4.5 Philipp Hummer (Designer, Founder of SPOTTERON Citizen Science App Platform)

Citizen Science is rooted deeply in user interaction, communication, and community building if done right. With the FStS project, multidisciplinary science and public engagement will meet our communication-rich mobile platform to enable participants to contribute not only observations, but also thoughts, ideas and qualitative feedback on ocean-related public art. On the SPOTTERON Citizen Science App Platform, our team has created a wide range of interactive options to enable people and science projects to connect and share information, comments, and also personal experiences for in-situ observations and data submissions. These communication and community tools are also an integrated part of the "ArtSpots" app toolkit (www.artspots.net), in which people contribute data about street art, graffiti or urban art culture, and also form a community of interest. The open participatory project will form the backbone of digital participation in the future developments of FStS project.

For the user's perspective, besides uploading a photo, an in-depth questionnaire stands ready about what impacts the direct observation of ocean-related art has for the viewer. By reflecting on the emotional spectrum and effect of art, a connection to the artwork, and its theme emerges, which combines the visual experience with an understanding of the importance and meaning. We acknowledge that everyone has a relationship to our seas and oceans, and to that extent also to ocean well-being. With that connection in mind, the FStS project serves as a true interdisciplinary and interactive hub that combines contemporary arts, social sciences and marine ecology in a modern mobile and digital-social environment.

8.4.6 *Marta Dieviņa (Aquatic Biologist, Science Educator)*

In my experience, I have noticed a trend—very often researchers create their inner information bubble in which they talk to each other in their own language of terms. As long as the researchers only communicate with each other, everything goes relatively smoothly. But when it becomes necessary to share the research topic to someone outside the inner circle, then the question arises—how to tell about it beyond the terms? And why are researchers "out of their boat" when speaking outside the scientific terminology? In my opinion, this is because academia might be too disconnected from society in a wider sense. In the past two decades, a branch of research that can reduce this disconnectedness has strengthened its place in the research field—it is citizen science. When I first heard about this research direction several years ago, I was amazed—how is it possible that everyone can help in research? Over the years, I have included citizen science more and more in my research activities, until I even became one of the creators of the FStS initiative. Since I am participating in this ocean literacy and street art related project, I have noticed that both my peers outside academia better understand the purpose of my work, and myself I am more capable to switch research terminology in public education activities. My experience shows that citizen science promotes understanding between different groups of people, which consequently helps to build and strengthen a more inclusive society in the long term.

8.4.7 *Natali Lazzari (Marine Conservation Scientist)*

FStS allowed me to witness first-hand how art can powerfully represent and reinforce the human-ocean connection, but it also highlighted the need to co-create inclusive narratives that resonate with diverse audiences.

While conducting project surveys in Galicia, I asked people about the emotions evoked by a series of murals. One mural in particular, featuring dark colours, struck me as conveying fear and sadness. Indeed, most people I surveyed shared this interpretation. However, one response surprised me. A grandfather, spending time with his grandson, told me that this same mural evoked happiness and hope for him. His answer caught me off guard, so I repeated my question to ensure he understood. After completing the survey, I struck up a conversation with him, asking why he associated a mural with such dark tones with happiness and hope. He explained, "It reminds me of the joy and hope I felt when my father and grandfather, who were artisanal fishers, returned home safely after a day of fishing during stormy weather".

This contrasting interpretation struck me, demonstrating that the same piece of art can evoke a wide range of emotions, from hope to fear. This experience helped me understand the importance of using transdisciplinary approaches that consider the ocean identity of the local community when creating art intended to foster specific social values, such as environmental awareness. Through FStS I learned that street art is a powerful medium for enhancing the human-ocean connection, but it also requires careful consideration of the diverse interpretations it can elicit.

8.4.8 Ludovica Montecchio (Master's Graduate)

When I joined the team of the project FStS, I worked as a translator for the surveys on murals, exploring seemingly clashing emotional responses to aesthetic, provocative, and enigmatic murals. The vast array of emotional responses provided allowed the observers to give multifaceted responses to these pieces of art, which were often ambivalent.

I believe that ambivalence is a core experience of humans' perception of the ocean. Hence it is a strength of qualitative transdisciplinary research. Ambivalence holds the unveiling power rooted in art's transformative potential: it moves. In times of ecological grief, all sorts of emotional responses play a key role as engines for conservation, even those normally defined as negative: eco-anxiety, worry, or grief can generate engagement through participatory approaches, as well as strategies that are place-based and culturally relevant (Ojala et al., 2021). The surveys showed that care can stem from emotional responses to art and to murals in particular, which are the most accessible existing form of art as they provide an educational experience that is not shared in institutional settings, but literally on the street. Murals generate an emotional response that empowers observers to become citizen scientists. It was moving to see how marine-themed murals can shape ocean identity through the combination of art, emotion, and science because our ocean identity is much like the motions of a wave, or those of an emotion—fluid and complex.

8.4.9 Svea Rogge (Master Student)

Deepening my studies in ecology and environmental planning, I started to favour interdisciplinary approaches in the field of conservation science and linguistics. In understanding and protecting our environment, we are urged to work together in every discipline, nation, language and living condition. Since most of our environment consists of and depends on the ocean, this field especially needs multifaceted approaches and multiple languages. English is not my native language and when I talk about my work I often flounder and mix up the words conservation and conversation. While laughing, slightly ashamed of my mistake, I realised that these terms are really close to each other in spelling and semantics. We need to have conversations in order to achieve conservation goals. The language of our conversation doesn't have to be word-based, it can be emotional, musical, artistic, and physical. Connecting multiple disciplines and people in more just and holistic conversations on conservation demands multiple language learning and an openness to all perspectives. Making "mistakes" in these languages must not be a deal-breaker in being part of the discourse but can enrich and deepen our understanding of each other. Diversifying our abilities of speaking, writing, painting, moving in different languages makes us more qualified to truly feel another perspective. Citizen science and arts incorporate aspects of this essential exchange. Bringing the sea to the streets of cities is a beautiful way of saying "let's have a conversation on marine conservation because it's connected to all of our lives even if we don't think about it every day"—it's just said more concisely with a spray can.

8.5 Theme Three: "From Sea to Street"—Lessons Learned Towards Building Ocean Identities

8.5.1 Milena A. Schreiber (Marine Researcher)

When I first came across the FStS project, I was immediately drawn to the idea of exploring how artistic practices reflect ocean spaces identities and not only represent, but also produce and enact different worlds. Throughout the life of the project, the co-creation of knowledge between academics and artists has led me to ponder this question: is this a process of co-creation, or a subtle way of influencing the 'free' minds of artists, hindering their own ways of articulating ocean identities? Could we be throwing the baby out with the bathwater? Murals can provide a wealth of knowledge about how artists see themselves in relation to marine spaces, but if murals begin to express academics' and artists' perceptions of how these relations should or should not be, the risk of losing this valuable source of knowledge may be high. The 'goods' of art as self-expression and interpretation could be discarded in an attempt to support certain values without the certainty of where those values come from. Legitimate researchers can carry authority even in participatory processes (Esguerra & van der Hel, 2021) and if the artists involved see the researchers as legitimate carriers of knowledge, we may exclude their perspectives. Any co-creation or transdisciplinary exercise requires monitoring and evaluation (Norström et al., 2020) and FStS could benefit from following this simple rule. I think it is worth starting a conversation about these considerations, not only as part of an academic discussion, but also because—to be honest—the best way for me to alleviate my own eco-anxiety is to unglue myself from my desk and visit a nearby art exhibition.

8.5.2 Marien Helmus (Master's Graduate)

During my interviews with mural artists, one recurring theme was how artistic freedom can be limited when work is based on commissions. One of the interviewed artists noted:

> "There's this scale that, on one end, you get paid a lot of money and the client tells you exactly what to do, and you are a slave of sorts. And there is the other side, where you don't get paid, but you get to do whatever you want. That's total freedom. There's a price to it".—Interviewee 3

This dynamic can impact the representation of ocean identity in murals, potentially reflecting the commissioner's vision rather than the community's. A way to circumvent this issue could be through the utilisation of co-creation processes. Co-creation processes can ensure the artwork aligns with the community's ocean identities. One artist emphasised the importance of engaging with the community:

> "If I leave a mural with people, I want it to be a grounded mural as well. So what I've done now in [mural location] for the community is an ode to that community. I met and talked to them, to feel more soul, to be able to create a work from a gut feeling that ultimately resonates with that same audience".—Interviewee 7

Additionally, by involving co-creators, dialogue about what to show and why certain things should be shown, could enhance the ocean identity of everybody involved in the process. This increase in dialogue about the ocean could furthermore help in enhancing ocean literacy. As Kelly et al. (2023) pointed out, ocean identity can play a valuable role in enhancing societal ocean literacy.

8.5.3 Rihards Bražinskis (Street Art Tour Guide)

As part of FStS, the main point of mine was to illuminate a quite paradoxical situation in Riga's graff'n' art scene. Namely—for some reason despite that Riga (and the whole country of Latvia) is endowed with a lot of coastlines and water reservoirs, the aquatic-themed pieces are practically absent. With maybe some exceptions of visual characters or potential of commission pieces. At the same time, the presence of water itself symbolically is very significant for Riga's graffiti and street art historical expansion. Starting with the 90ies Maskachka tram tunnel as a cradle of it all and continuing with Andrejsala as an unrepeatable blast of alternative cultural activities during the late 2000'—both located next to river Daugava. Neglected spots next to some water—under the bridges, military legacy concrete fences in bushes, or sad rusty metal sheds in almost countryside like areas—is a natural habitat where it all blooms on it's own as a blurred borderline between the urban and the rural. A little bit different is the case study of Liepāja city next to the Baltic Sea, where the marine-themed pieces are more common and even more integrated in the city. In Liepāja it's more natural and romantic to have this kind of work, as we can call this town as a place of resort. There this type of vibe is more organic.

8.5.4 Cika Annemae (Street Artist)

Within the street art atmosphere in Utrecht, many of the murals show reference to nature. A few examples in Utrecht are; a giraffe interplaying with the architecture of the building, otters playing in the cartoonish Utrecht setting, godwits, and other local birds. As a muralist myself, I love to paint animals, insects, and organic forms (Picture 8.3). To me, it stands symbolically for our ultimate connection with our natural surroundings. As humans, we are part of (and no better than) the natural world around. Painting is a way of paying respect and giving attention to the things that can be overlooked in city life. Challenging for example, the stereotypical beliefs that insects are dirty and annoying and can be killed without a moment of doubt.

Picture 8.3 A mural (2022) in Willemstad, Curacao by Cika Annemae. Copyright: with permission by the artist

Street art can contribute to the appreciation and awareness of nature conservation. Having worked as a tour guide in galleries and museums—SAMA being one of them—I have found that all kinds of people, from kids to corporate workers, are well able to read the often layered meanings of a work of art. The only necessity for understanding the message is a moment for consideration and sometimes the right questions (Visual Thinking Strategies for example, provides a usable frame set).

Art in public space can trigger people into these moments of contemplation when they are passing by. It functions almost like a refreshing opposite to advertisement. But the message can also be easily overlooked or ignored, or even function as a welcoming decoy that makes you smile instead of fixing the problem at stake. If you look at deeds rather than ideas, one must admit that painting walls about nature, does in no way provide more space for nature itself. It will even do some of the opposite. Outside walls are often home to spiders. By using spray paint, I killed more than I had wished.

8.5.5 Sophia Kochalski (Social Fisheries Scientist)

Through FStS, I was able to better understand the northwest of Spain in terms of its collective ocean identity, defined as a regional identity "rooted in how the ocean as a place supports the sense of self" (Buchan et al., 2024). As I learned, many artists do research before creating a mural and incorporate local elements into their art. Therefore, murals often reflect typical elements of the local culture.

In the Galician town of Redondela, for example, murals showcase cuttlefish, the town's most famous dish, as well as the legendary local sea monster "la Coca". Likewise, many murals in Galicia illustrate themes related to fishing, traditional agricultural activities, native plants, or notable regional artists. On a more abstract level, several murals convey a sense of melancholy or nostalgia typical of Galicia, known as "morriña". However, as our FStS surveys and interviews revealed, such interpretations depend on the cultural knowledge and background of the observer.

Based on the murals, it appears that many Galicians have a strong ocean identity. At the same time, the sea is part of a broader regional identity that includes both marine (María, 1991) and non-marine themes (Miguélez-Carballeira, 2014). Occasionally, murals spark controversy, particularly when they feature motifs related to demands for greater regional independence. This shows that murals not only reflect identity but also contribute to the complex process of negotiating regional identity.

8.6 Key Points and Lessons Learned

Based on our experiences and personal reflections, authors from diverse backgrounds shared valuable insights in this chapter that highlight the significant role of street art, particularly murals, in promoting ocean conservation and strengthening community engagement:

(1) Murals were presented as powerful educational tools that make complex environmental issues like overfishing, plastic pollution, and climate change accessible and enhance ocean literacy. Their emotional impact can inspire people to actively engage in ocean conservation. The integration of citizen science with street art enhances this effect, enabling individuals and groups to participate in conservation projects and gain a deeper understanding of ocean-related issues.

(2) Street art collaborations were suggested to democratise art by amplifying diverse voices and integrating various knowledge systems. Such projects highlight the importance of cultural relevance and co-creation in ensuring that murals truly reach the communities they are intended to address. This enriches our understanding of the relationships between humans and the ocean and shows how art can act as a catalyst for social change.

(3) Murals provide a platform to strengthen communities and foster dialogue. They are particularly effective when culturally relevant and aligned with the values and surroundings of local communities. Street art uniquely connects people with the ocean, its creatures, and the communities tied to it, deepening awareness of marine conservation.

A recurring theme was also that street art is much more than mere aesthetic decoration. Based on our experiences, embodied here as personal reflections, we argue that ocean-related street art creates space, community and identity, and supports citizens' engagement with the ocean environment and culture. Balancing artistic freedom with co-creation is crucial to ensuring that murals remain both authentic and impactful.

8.7 Concluding Remarks

In our reflections, we emphasised how street art, especially ocean-related murals, has opened up new perspectives and allowed us to reflect on unexplored elements of our own lives, surroundings and work. For example the artists emphasised the care taken in the documentation, research process and in painting the murals, while the researchers and educators reflected on how street art complements and supports their practices, broadening the mindset and creativity. From our shared experiences, it emerged that ongoing street art initiatives have the potential to initiate art-science collaborations and through these collaborations, art can be democratised and people can be (re)connected with the ocean. Looking ahead, interdisciplinary approaches that combine art, science, and community engagement offer a promising path to advance marine conservation. By continuing to explore innovative ways to integrate these elements, we can create more inclusive and culturally relevant initiatives that foster the integration of diverse knowledge systems and a deeper connection to the ocean.

Acknowledgements We appreciate all the input from the project participants as well as our project partners. IMPETUS supported the "From Sea to Street" project and is funded by the European Union's Horizon Europe research and innovation programme under grant agreement No. 101058677. Arias Schreiber was supported by the European Union's Horizon 2020 Research and Innovation Programme under the Marie Skłodowska-Curie grant agreement No. 843165. Natali Lazzari has been funded by the European Union, MSCA program, REVALSEA, No. 101146373. Sebastian Villasante gratefully acknowledges the financial support from EQUALSEA (Transformative adaptation towards ocean equity) project, under the European Horizon 2020 Program, ERC Consolidator (grant agreement No. 101002784) funded by the European Research Council.

References

ARTigo. (n.d.). *ARTigo: Social image tagging*. https://ars.electronica.art/citizenscience/en/artigo-social-image-tagging/

Bonhoure, I., Cigarini, A., Perelló, J., & Vicens, J. (2019). *Citizen science and the arts*. https://lab.cccb.org/en/citizen-science-and-the-arts/

Buchan, P. M., Glithero, L. D., McKinley, E., Strand, M., Champion, G., Kochalski, S., Velentza, K., Praptiwi, R. A., Jung, J., Márquez, M. C., Marra, M. V., Abels, L. M., Neilson, A. L., Spavieri, J., Whittey, K. E., Samuel, M. M., Hale, R., Čermák, A., Whyte, D., ... Payne, D. L. (2024). A transdisciplinary co-conceptualisation of marine identity. *People and Nature*, 6, 2300–2324. https://doi.org/10.1002/pan3.10715

DITOs consortium. (2019). *Citizen science and art/science: Synergies and future potential* (DITOs policy brief 10).

Eco-arts project. (n.d.). *RECOVERY*. https://www.bmwhi.org/recovery-creative-arts

Eisner, W. E. (2003). On the differences between scientific and artistic approaches to qualitative research. *Visual Arts Research, 29*(57), 5–11.

Esguerra, A., & van der Hel, S. (2021). Participatory designs and epistemic authority in knowledge platforms for sustainability. *Global Environmental Politics, 21*, 130–151. https://doi.org/10.1162/glep_a_00573

focusTerra. (2020). *focus Terra*. https://reisesteine.ethz.ch/en/about

Heras, M., Galafassi, D., Oteros-Rozas, E., Ravera, F., Berraquero-Díaz, L., & Ruiz-Mallén, I. (2021). Realising potentials for arts-based sustainability science. *Sustainability Science, 16*, 1875–1889. https://doi.org/10.1007/s11625-021-01002-0

Huynh, C., & Jürgens, A. S. (2024). Christina Huynh: The Charisma of Nature in Street Art. *w/k–Between Science & Art Journal*. https://between-science-and-art.com/christina-huynh-the-charisma-of-nature-in-street-art/

Kelly, M. R., Kasinak, J. M., McKinley, E., McLaughlin, C., Vaudrey, J. M., & Mattei, J. H. (2023). Conceptualizing the construct of ocean identity. *npj Ocean Sustain, 2*. https://doi.org/10.1038/s44183-023-00025-7

María, M. (1991). A presencia do mar na poesía galega. *Revista Internacional de los Estudios Vascos/Nazioarteko Eusko Ikaskuntzen Aldizkaria/Revue Internationale des Etudes Basques, 36*, 313–338.

Miguélez-Carballeira, H. (2014). A companion to Galician culture (Vol. 344). Boydell & Brewer Ltd.

Norström, A. V., Cvitanovic, C., Löf, M. F., West, S., Wyborn, C., Balvanera, P., Bednarek, A. T., Bennett, E. M., Biggs, R., de Bremond, A., & Campbell, B. M. (2020). Principles for knowledge co-production in sustainability research. *Nature Sustainability, 3*. https://doi.org/10.1038/s41893-019-0448-2

Ojala, M., Cunsolo, A., Ogunbode, C. A., & Middleton, J. (2021). Anxiety, worry, and grief in a time of environmental and climate crisis: A narrative review. *Annual Review of Environment and Resources, 46*, 35–58. https://doi.org/10.1146/annurev-environ-012220-022716

Prūse, B., Kochalski, S., Dieviņa, M., Gianelli, I., Roumbedakis, K., Helmus, M., Ekström, E., Montecchio, L., Straka, T., Rogge, S., & Kinds, A. (2024). "From Sea to Street": Initiating change for stronger connection with our ocean. ECSA "Change—The transformative power of citizen science", Vienna, Austria, Arpha-System of Pensoft. https://doi.org/10.3897/ap.e126595

Riggle, N. A. (2010). Street art: The transfiguration of the commonplaces. *Journal of Aesthetics and Art Criticism, 68*(3), 243–257.

Rüfenacht, S., Woods, T., Agnello, G., Gold, M., Hummer, P., Land-Zandstra, A., Sieber, A., Vohland, K., Land-Zandstra, A., Ceccaroni, L., & Lemmens, R. (2021). Communication and dissemination in citizen science. In K. Vohland et al. (Eds.), *The science of citizen science*. Springer.

Sanchez, E., Vinueza, R., Izurieta, X., & Rey, N. (2020). Use of muralism to promote awareness about aquatic ecosystems and wise water consumption in northwestern Ecuador. *Ocean & Coastal Management, 190*, 105165. https://doi.org/10.1016/j.ocecoaman.2020.105165

Schneller, A. J., & Irizarry, A. (2014). Imaging conservation: Sea turtle murals and their effect on community pro-environmental attitudes in Baja California Sur, Mexico. *Ocean & Coastal Management, 89*, https://doi.org/10.1016/j.ocecoaman.2013.12.009

Sea Walls: Artists for Oceans. (n.d.). *About us*. https://seawalls.org/about/

Smaniotto, A. (2023). The social sciences and humanities contributions to citizen science: insights from the COESO project journey (COESO D.7.6). Zenodo.

Tauginienė, L., Butkevičienė, E., Vohland, K., Heinisch, B., Daskolia, M., Suškevičs, M., Portela, M., Balázs, B., & Prūse, B. (2020). Citizen science in the social sciences and humanities: The power of interdisciplinarity. *Palgrave Communications, 6*. https://doi.org/10.1057/s41599-020-0471-y

Tyng, C. M., Amin, H. U., Saad, M. N. M., Malik, A. S. (2017). The influences of emotion on learning and memory. *Frontiers in Psychology, 8*. https://doi.org/10.3389/fpsyg.2017.01454

Worm, B., Elliff, C., Fonseca, J. G., Gell, F. R., Serra-Gonçalves, C., Helder, N. K., Murray, K., Peckham, H., Prelovec, L., & Sink, K. (2021). Making ocean literacy inclusive and accessible. *Ethics Science and Environment Politics, 21*. https://doi.org/10.3354/esep00196

CHAPTER 9

The Power of Music to Foster Ocean Stewardship: An Experience from the West Coast of Ireland

Maria Vittoria Marra and Seán McGrath

9.1 Introduction

Why music? Because we are in Ireland, is the answer. Of all the images that come to mind when thinking of the "Emerald Isle", music sessions in pubs are probably amongst the most common. And that is not far from reality. Especially on the West coast of Ireland, it is absolutely normal to step into a pub and find a traditional Irish music session on. Fiddles, concertinas, whistles, mandolins, banjos, guitars, accordions would all be common instruments at these sessions. Musicians seem to be able to join ongoing sessions with absolute ease, they do not need to have music sheets with them, they are just able to play "*by ear*" (Kearney, 2024), or can join with a song, if they wish. In reality, the dynamics of these

M. V. Marra (✉)
Galway Atlantaquaria, Galway, Ireland
e-mail: maria@nationalaquarium.ie

S. McGrath
Galway, Ireland

C. Certomà (ed.), *Blue Kinships*,
https://doi.org/10.1007/978-3-031-78619-8_9

traditional music sessions in Ireland are probably more complex than the romanticised view of tourists and visitors (O'Shea, 2007), but it is undoubtedly true that music is one of the foundations of Irish culture and it attracts musicians from all over the world to this small island (Rapuano, 2001). This profound relationship between music and Irish people is manifested in the fact that in many settings—especially rural—it is common for children to start learning one or more instruments. Thus, people are exposed from a very early learning stage to the broad repertoire of Irish tunes. These are normally made of patterns of short musical phrases and are based on the concept of repetition of each of their parts (Kelly et al., 2010), allowing them to be learned quickly by ear. In such a context, where music is almost treated like an additional language commonly spoken (Keegan, 2011), it became natural to consider that the incredible bonding potential of music could be harvested to enhance people's connection with the ocean.

Galway presented itself as a perfect location for this experiment. In fact, in spite of its relatively small size, this city is considered to be the cultural capital of Ireland. In addition to hosting two Universities, it is the venue of at least 20 different cultural festivals,[1] in particular the International Arts Festival, which takes place each July in Galway City. Founded in 1978, it covers a range of art forms.[2] In this incredibly fertile environment from a cultural perspective, individual initiatives proposed by private citizens can go a long way in giving access to music to anyone. Three successful examples that have met massive growth in the last ten years are:

- The "Galway Ukulele Group", a joyful gathering of enthusiasts of the "*diminuitive four-stringed guitar*" (Tranquada & King, 2012) which has become the beating heart of the increasingly popular international "Galway Uke Fest"[3];

[1] https://thisisgalway.ie/galway-festivals-galways-events-and-festival-calendar/, accessed on 31 July 2024.

[2] https://www.giaf.ie/info/about, accessed on 31 July 2024.

[3] https://galwayukefest.com/, accessed on 30 August 2024.

- The "Slow Trad Session",[4] where Irish tunes, normally played at very high pace, are slowed down dramatically to allow learners to practice them in a group setting and become familiar with the "ornamentations", i.e. the different stylistic determinants characterising each of the numerous regional styles (McCullough, 1977).
- The "Galway Jam Circle",[5] in which musicians play in a circle formation simple three- or four-chords songs which allow everyone to join easily and gives the same musicians the opportunity to improvise instrumental solos.

In spite of being very different in terms of approaches, dynamics, genres considered and instruments played, these groups have all two main elements in common. First, they offer free access to anyone willing to play music with others, regardless of their level of training and ability—In fact absolute beginners can join and learn to play an instrument from scratch! Secondly, and just as important, they promote the development of a sense of community founded on the love of the music.

9.2 "*Amhráin na Mara*", the First Project

It was this commonality between the community music groups in Galway that inspired the start of an experiment at the end of November 2020. This was the time period when the Northern hemisphere was experiencing the reinforcement of the lockdown restrictions due to the winter surge of Covid-19 cases. At the time, many group activities were still mainly done as virtual gatherings on the Zoom platform (Zoom Video Communications Inc, 2016). It was not unusual even for musicians to use this and other online platforms to find innovative ways to overcome the limitations on group gatherings caused by the pandemic. The number of videos showing home performances of popular artists (see Spice Girl Mel C and KT Tunstall's bathroom sessions[6]) increased steadily, and even amateur

[4] https://www.facebook.com/groups/329713190883326/, accessed on 30 August 2024.

[5] https://www.facebook.com/groups/galwayjamcircle/, accessed on 30 August 2024.

[6] https://www.independent.ie/entertainment/music/mel-C-and-kt-tunstall-to-perform-from-their-bathrooms-for-wateraid/39186770.html, accessed on 31 July 2024.

musicians became proficient users of softwares like Bandlab[7] which allow combining parts of recordings as in a professional music studio to work together on a music composition, despite the physical distance. In this context, the author of this chapter Maria Vittoria Marra (soon to become Education and Public Engagement Officer at Galway Atlantaquaria after that) noticed a low number of presentations about how music can convey ocean literacy while attending the marine science communication conference CommOCEAN 2020 (Cox et al., 2021). This observation resulted in a proposal to the co-author Seán McGrath, songwriter and founder of the Galway Jam Circle, and also to another group of Galway-based musicians called "The Galway Ukers" (of which Maria Vittoria is a member) to try to write a song about the ocean together, with the intent to raise awareness about it.

The enthusiastic response by Seán and the other musicians culminated in a project which was originally called "*Amhrán na Mara*" ("Song of the Sea" in Irish Gaelic), in honour of a popular Irish animated movie from 2014.[8] However, the name was soon changed to the plural "*Amhráin na Mara*" (AnM) because not just one, but multiple songs were produced as result of it, many of them co-authored by two or more of the project participants (Picture. 9.1).

A few of the songs developed within the AnM project focus on the threats to the ocean posed by pollution. Of them all, two in particular stand out due to their original perspectives. Both were the results of the collaboration between two musicians.

The first one is called "*Pigfish*",[9] a simple song form based on what is known by musicians as a "three-chord trick" with an ironic approach and a catchy chorus:

> I am the pigfish baby and the ocean is my home
> Through endless seas of garbage I love to roam
> Thank you thank you for making my world Kodachrome

The song was inspired by the character named Oscar the Grouch in the timeless Canadian TV programme Sesame Street, who happily lives in a garbage can and sings a song called "*I love trash*" quoting "*Oh, I love*

[7] https://www.bandlab.com/, accessed on 23 August 2024.

[8] https://www.youtube.com/watch?v=jStBiU_FhUI, accessed on 29 August 2024.

[9] https://www.youtube.com/watch?v=djxbcgefrME, accessed on 29 August 2024.

Picture 9.1 Seán McGrath and some Galway Jam Circle members performing a few of the songs developed within the AnM project in 2022 at Galway Atlantaquaria. Since that year, the aquarium has started hosting events featuring songs and music installations about the ocean as part of the programme of Culture Night[10] in Galway City. (Photo Garry Kendellen)

trash, anything dirty, or dingy, or dusty, anything ragged, or rotten, or rusty. Yes, I love trash!".[11] This concept was basically translated into the idea of a fish living in the ocean that loves all garbage and pollution dumped there. The name was chosen because pigs tend to be generally considered as dirty animals, but the authors were surprised to find out that there is an actual species named the pigfish, i. e. *Orthopristis chrysoptera* (Ohs et al., 2011). The lyrics of the verses were inspired by pictures found online of marine animals affected by all that litter that makes the pigfish's world "Kodachrome". This reference to the first colour film materials used in

[10] https://culturenight.ie/about/, accessed on 1 August 2024.

[11] https://www.youtube.com/watch?v=rxgWHzMvXOY, accessed on 20 August 2024.

cinematography and photography winks to the homonymous song by Paul Simon from 1973.[12]

Nothing I like more than living in a chemical spill
With 6-pack plastic that turtle is dressed to kill
Can you send me some more I just cannot get my fill

I have a sea horse friend. He married a Q-Tip dame
My octopus neighbour—a bottle top is his flame
Please send me some more running low would be such a shame

Oil on a bird's wing. A wonderful sight to see
Uranium in seaweed or maybe just mercury
I am begging for more I am the pigfish can't you see

Similarly, the song "*Barbie Gals*" is a visionary collaboration between two songwriters who have imagined the destiny of the popular dolls known as Barbie ending up gathering at the bottom of the ocean after being dumped. The song starts with a chorus that is hard not to sing along to:

Ooh arh, party on,
Ooh ooh arrh, come sing our song
Ooh arh, party on,
We will play with you.

Again inspired by images found on the Internet of dolls washed up on the shore as spooky contributors to its litter pollution, in this song, Barbies are represented in the guise of the sirens from the ancient Greek mythology who attracted seafarers to the coast with their melodious singing, often luring them into crashing their ships into rocks. Similarly, the Barbies gathered at the bottom of the ocean, almost part of a separate underwater universe reminiscent of Verne's "*Twenty Thousand Leagues Under the Sea*" (1916), call everyone to join their song of disbelief about global warming and the consequences of oblivious consumption, attracting society towards a fatal sinking destiny. The contrast between

[12] https://www.paulsimon.com/track/kodachrome-7/, accessed on 27 August 2024.

the uplifting rhythm of "Barbie Gals" and its unsettling lyrics is further highlighted by the choice of starting the song with the very popular vocal harmony arpeggio "borrowed" by famous tunes like the Beatles' cover "*Twist & Shout*"[13] and David Bowie's "*Let's Dance*".[14]

> We will tell you something you might like to know
> Have you ever wondered where old Barbies go?
> They sink to the bottom of the sea
> Come and join our party.
>
> We are the Barbie Gals we will be your pals
> But don't mess with us or make a fuss
> Just do what we say
> Never reuse what you can throw away
>
> We have the power to control the news
> Making you believe just what we choose
> Global warming is just fine
>
> You can consume more every day
> Nothing left but money one fine day
> Reckless consumption is your friend
> Eat the money you can't spend

For other songs developed within AnM, the sources of inspiration were the personal experiences of the musicians and their own connections with the ocean. This is not surprising, consider that the songwriting process is considered to be a special place for connection with oneself and the inner world (Beech, 2015). Also, this is particularly relevant in the context of this chapter considering how emotional connections to the ocean, or "*Emoceans*", have been proposed by Emma McKinley et al. (2023) as one of the new dimensions to include in the ocean literacy framework, given

[13] https://open.spotify.com/track/5ZBeML7Lf3FMEVviTyvi8l, accessed on 20 August 2024.

[14] https://open.spotify.com/track/3ix6K4wZY29bCujrSznwFZ, accessed on 20 August 2024.

Plastic degrades into wine.

the important role they can play in driving behaviour change in favour of ocean conservation and action. In this sense, an example from the AnM set of songs is "*Morning swim*" which was inspired by the habit of the songwriter's wife and her group of fellow female swimmers, who get into the water in Salthill (Galway) all year round to jolt their bodies and get energised by their morning swim. The practice of swimming in cold waters, regardless of the time of the day or year, is increasingly considered beneficial for both physical and mental health (Whitworth, 2018). That was particularly evident at the time when this song was written, i.e. during the Covid-19 pandemic, when the number of ocean swimmers in Ireland increased exponentially (Foley, 2022).

The sensory experience of the enveloping materiality of the ocean involved in the swimming activity, as well as the sense of wellbeing received through it, are all examples of what has been defined by Buchan et al. (2024) as "*thalassophilia*", considered as "*the product of strong emotional responses to the sensorial experience of the ocean and shared social or cultural understanding of ocean place identifications*". This concept can be found expressed also by the individual experiences described in two additional songs written for the AnM project, one is "*Teal Edges*". This chorus-less song describes how the author of its lyrics learned to explore the world within herself while walking along different shorelines, first in the place where she was born, then in the place where she chose to live, to climax in the end with the absolute certainty that, wherever she can find these "teal edges", she will be able to feel safe enough to continue becoming who she is. The other song is called "*My baby blue*", and it was inspired by the songwriter's reflections on the years spent living in the US and all the times flying between America and Ireland over the ocean. The thought that the ocean is this massively large space we fly over during overseas trips without being able to see it from above the clouds made him consider how, compared to the times when people used to traverse it by ship, the ocean is becoming an increasingly abstract concept for many people, especially the ones living far away from it. In this sense, he considers the Irish population, 40% of whom live within 5 km from the coast,[15] extremely lucky to maintain this connection with the ocean: he says "*I always know where the ocean is, and that makes me feel grounded. It*

[15] https://www.cso.ie/en/csolatestnews/pressreleases/2017pressreleases/pressstatementcensus2016resultsprofile2-populationdistributionandmovements/, accessed on 13 August 2024.

is a kind of compass for me, and I feel I'm oriented in space because I know where the ocean is". The seed for this song was the consideration that the word "blue" tends to be used in songwriting's contexts of sadness or melancholy. In fact, there is an entire music genre, the Blues, which originated by the mixing of African and European musical traditions when slaves were brought into America, and it is characterised by "out of tune" sounds generated by bending the twelve notes of the traditional Western music system on guitars and harmonicas (Gibson, 2009), so generating the deep emotions conveyed by Blues music. The songwriter's intention for the AnM song "*My baby blue*" was to turn away from the perception of the word "blue" as associated to a negative emotional state, to focus instead on the blue as the colour of the ocean, to compose what essentially is a love song expressing worry for the baby (i.e. the ocean) that is not happy and is crying salty tears.

The above is only a selection of the twelve songs that were developed within the AnM project, but it is interesting to note how, despite the differences of genres and approaches considered, the ocean themes selected as subjects for their songs by the musicians involved reflect the results of an ocean literacy survey carried out in Ireland within the framework of the Fair Seas campaign.[16] This survey, deployed to investigate the existing levels of ocean literacy of the Irish people and their relationships with the ocean, found that 60% of survey respondents feel awe/wonder when they think about the ocean, while for 47%, the main emotion is concern about the marine environment (McKinley, 2023). From this latter perspective, the survey commissioned by Fair Seas showed how marine litter and plastic pollution are more widely considered as threats than overfishing, climate change and habitats alteration, similarly to what Lotze et al. (2018) had already demonstrated from a global perspective.

9.3 The Project with Claddagh National School

The evidence that marine litter is the main concern associated with the ocean was furtherly confirmed by a follow-up project developed by Galway Atlantaquaria in collaboration with Seán and The Galway Ukers in 2022 as part of the outreach activities the aquarium carries on in primary schools for the Marine Institute's Explorers Education

[16] https://fairseas.ie/about/, accessed on 22 August 2024.

Programme.[17] One of these is the special all-school project called "*Ocean Champions Awards*", for which entire schools develop a project based on how students imagine a healthy and clean ocean to be. In 2022, Galway Atlantaquaria worked with Claddagh National School in Galway, which has the pretty unique characteristic for a school of owning a recording studio! As part of their Ocean Champions Awards project, the students in this school were first introduced to the ocean literacy principles and marine biodiversity. A particular focus of their study during the project was the squid, which the kids had also the opportunity to dissect. After these preliminary activities, as an assignment, the students developed a podcast[18] about their favourite sea species along with their teacher, Mr Seán Leonard, who is also a theatre actor and a local city guide. This brilliant and innovative project prompted the proposal by Maria Vittoria for the school to work on a music project aimed at the co-development of a new original song. The idea was for the kids to reflect on what they had learned during the time spent working on the Explorers project, and their own relationship with the ocean, and then write down all their thoughts, dreams, and ideas about how a healthy ocean should look like in the form of short poems. Their compositions generally highlighted positive attitudes towards the ocean and sense of care, as well as the awareness of the many human behaviours which should be reconsidered in order to reduce our impact on the ocean, aligning with what was already observed with slightly older youth in other parts of the world (Koulouri et al., 2022; Wootton et al., 2024). All the material produced by the students was then passed to Seán to inspire him in the songwriting process aimed at creating a song suitable to be sung by the students along with the same Seán and the Galway Ukers.

A particular line suggested by one of the students caught Seán's attention, namely "*Great Pacific garbage patch removed*". As a musician, Seán was immediately drawn to those words because he thought "Great Pacific Garbage Patch" had already got a rhythm to it. He had never heard the phrase before, and had to google it to find out that it refers to a massive accumulation of buoyant plastic litter created in the North Pacific by the ocean circulation (Lebreton et al., 2018) as big as twice the size of

[17] www.explorers.ie, accessed on 19 August 2024.

[18] https://www.claddaghns.ie/claddagh-ns-news/2022/4/29/the-podcast-is-back-claddagh-kids-talk-the-ocean, accessed on 18 August 2024.

Texas.[19] This discovery was so shocking for Seán that he decided to call the song "*The Great Pacific Garbage Patch*". To compose the lyrics, he both picked some of the topics mentioned by the kids that had grabbed his imagination, as e.g. the harmless looking, but actually non-degradable, baby wipes, but also decided to add some additional pieces of information he had researched himself. Despite the "*horror*" described in the lyrics (as defined by Seán), he decided to use a simple and sweet piano arrangement suitable even for young children to sing.

In the Great Pacific Ocean a wonder can be found
Made by me and made by you it's a million miles around
The Great Pacific Garbage Patch is a horror best unseen
How I wish the ocean was clean was clean
Oh how I wish the ocean was clean

We throw away our plastic stuff not knowing where it goes
Someone makes it go away somehow, I suppose
What harm is a baby wipe, an elastic band or two?
How I wish the ocean could be blue could be blue
Oh how I wish the ocean could be blue

In a school beside the ocean we see the waters rise
And the science tells us clearly that this is no surprise
Pollution is not free of cost it will kill the sea someday
Oh how I wish the fish were here to stay to stay
Oh how I wish the fish were here to stay

So keep the beaches clean use plastics less and less
Treat the ocean like a friend filling you with happiness
Bit by bit we all can change, a little every day

And make that garbage patch go away go away
We can make that garbage patch go away.

[19] https://theoceancleanup.com/great-pacific-garbage-patch/, accessed on 19 August 2024.

After composing it, Seán played the song first to the members of The Galway Ukers who had decided to take part in this new project. One of them suggested making the song more personal for the kids by replacing the line "*In a school beside the ocean we see the waters rise*" with "*In Claddagh school on Galway Bay we see the waters rise*". At this stage, the song was shared with the students and their teachers, including the music teacher who took charge of teaching the song to the kids and preparing them to meet the musicians to perform the song together in the recording studio of the school. A video of the song,[20] renamed afterwards by the students "*How We Wish The Ocean Could Be Blue*", was produced combining pictures from the day of the recording (Picture. 9.2) and marine pollution drawings made by the same kids. When asked in a survey put together by Maria Vittoria what was their feedback about the music project, the majority of the students said they thoroughly enjoyed it, but they preferred to dissect the squid!

Picture 9.2 The musicians and the students practising the song together in the recording studio of Claddagh National School (Photo Seán Leonard)

[20] https://www.youtube.com/watch?v=ARNQ4pNb4Dg, accessed on 22 August 2024.

It is interesting to note that the number of Galway Ukers who took part in this second music project with Galway Atlantaquaria was reduced compared to AnM. No specific analysis was carried out to find out why some project participants dropped after the first project, but informal conversations pointed towards the big role that the confinement due to the pandemic played the first time. In fact, AnM largely took place during the winter of 2021, when the repeated restrictions imposed by the lockdown were tough for many, therefore the participation in the project felt like a nice opportunity to do something different and to still interact with fellow musicians virtually. In this regard, it is worth reminding that some of the songs produced for AnM were co-developed by two or more musicians, and that for many of them, it was the first time they had ever written a song.

9.4 "Leave No Trace"

However, the enthusiasm shown by some of the musicians through the two projects encouraged Maria Vittoria to embark on a new adventure in 2022, namely a project proposal in collaboration with Music for Galway (MfG). This is a leading organisation in Ireland for the promotion of classical music who over the years have increasingly committed to promote sustainability in classical music settings. This culminated in 2023 with the funding of an international Creative Europe project called "*Songs of Travel*",[21] aimed at raising awareness and fostering empathy for the migrations caused by climate change through music. Therefore, it was not surprising for the CEO of MfG, Anna Lardi, to enthusiastically decide to work with Galway Atlantaquaria on the development of a project proposal aimed at increasing the ocean literacy of local musicians and inspire in them behaviour change and climate action. The project was called "*Galway Bay is Calling*" (GBiC) to stress the urgency of the intervention required as well as to pivot on the emotional connections people living in Galway have with the bay, and it was funded in 2023 by the Creative Climate Action Strand of the Creative Ireland Programme.[22] GBiC was conceived as a participatory art project engaging three community music groups based in Galway via their attendance in a series of marine science

[21] https://www.songsoftravel.eu/, accessed on 20 August 2024.

[22] https://www.creativeireland.gov.ie/en/about/, accessed on 20 August 2024.

and behaviour change workshops. Between September and November 2023, the project participants joined a rockpooling experience aimed at introducing them to the life on the shore to overcome the perception of the ocean as an alien and distant world; afterwards, the singers and musicians were taken on a journey to discover the seven principles of ocean literacy through an interactive tour of the aquarium. Finally, they had the opportunity to learn about the impacts of their everyday choices from plastic use to e-waste, and what they can do to mitigate those, through their participation in a dedicated behaviour change workshop led by Dr Joao Frias (Atlantic Technological University). The same workshops were also attended by the cello composer and songwriter Naomi Berrill, who had been involved in the project to create an original music composition inspired by her own learnings as well as the experiences of the other project participants. To do so, Naomi met individually with the three ensembles during their own usual rehearsal times to discuss everyone's main take-aways from the ocean literacy workshops and how these could possibly be translated into music (Picture. 9.3).

Born in Ireland, but based in Italy, Naomi Berrill is an internationally acclaimed composer who has repeatedly reflected on her own ocean connections in the music she has produced. A clear example of this is her

Picture 9.3 A moment of the workshop with the musicians of the Galway Jam Circle led by Naomi Berrill in November 2023 (Photo Garry Kendellen)

last album called "*Inish*" (Irish for "island", 2023), a celebration through music of the wonders and vicissitudes of life on the islands of Inishark and Inishbofin off the West coast of Ireland.[23] After meeting the musicians involved in GBiC, Naomi went on to compose the music piece inspired by this co-development experience. This creative process took over four months and it culminated in the genesis of "*Leave No Trace*", a fifteen-minutes otherworldly composition for cello and "*mixed bag orchestra*", as it was appropriately defined by Anna Lardi to stress the very diversified nature of the groups of amateur musicians and singers involved in GBiC.

These were a string orchestra (the "Galway Camerata"), classically trained and accustomed to reading music scores, a choir also accustomed to arranged performances and to be led by a conductor (the "Voice of Galway") and, on the opposite side, the already mentioned Galway Jam Circle, mainly comprising self-trained musicians more used to improvise in so-called "jamming" performances rather than playing pre-determined arrangements based on reading scores/music sheets, and so lacking an actual conductor in favour of its founder who acts more like a facilitator. The diversity of the three ensembles represented a remarkable challenge for Naomi Berrill, who paid all possible care to incorporate into the music composition she created the different styles, levels of music training as well as the learnings and the perspectives about the ocean of the project participants.

Naomi's final piece testifies to the potential that music has to tell stories without using words and to greatly amplify the effect of a small number of words. In fact, through the variations of its mood, "*Leave No Trace*" brings us through an emotive journey that alternates between concern/worry on one hand and optimism/hope on the other. In fact, the instrumental music played by the string orchestra creates the pathos that reflects the concern for the challenges climate change makes us face. When the composition moves towards an Irish/Celtic rhythmic form known as a "slip jig", it is to understand that we are in Ireland, and what we are concerned about is Galway Bay, above which "*my soul to soar forever more*". In the finale, with a bright, so-called "one four five" chord progression—a staple in popular music—the composition summarises the purpose of the whole project. The lines "*we are music makers and we are dreamers of dreams, we are rocks and shakers of the world forever it seems*"

[23] https://open.spotify.com/album/5Vvv7IyEFetEJb9crRuH8S, accessed on 21 August 2024.

describe the hope that the experience could transform the singers and musicians involved in the project into agents of change who will transmit their learnings and their experience to others in the community.

The original music composition was premiered by Naomi and the three ensembles on the 18th of May in what it was the first concert of CELLISSIMO 2024[24] a unique international classical music festival dedicated to the cello organised by MfG. The performance was very well received by the audience and the media, but probably most importantly, it was a very moving experience for the project participants. In fact, after practising their own parts for two months in their own weekly meetups, it was only when they attended the rehearsal the week before the premiere of "Leave No Trace" that they could see and hear how their experiences and contributions had come together to create the complete piece. The evaluation process involved in GBiC, which had the entire group of project participants, or samples of it, involved in perceptions surveys, a focus group and one interview (Frias et al., unpublished data) can only be completed after the end of the project in September 2024, therefore at a later time than this chapter is written, but it will help measure if the participation in the project has been effective in transforming the musicians and singers involved in ocean stewards. For now, it can only be reported how, during the week leading to the premiere of "*Leave No Trace*" and after the actual performance, the majority of the project participants were moved and unified by feelings of gratefulness towards Naomi, for her ability to create a stunning music piece in which they all felt like they have their part, and towards her brother and conductor Matthew Berrill, who masterfully brought the three ensembles and Naomi together with his kind leadership.

9.5 Conclusion

What is next, then? The unifying successful element for all the projects carried out from AnM to GBiC seems to be the songwriting process, described by Cobb (2016) to have the unique power to merge the emotional properties of music and language. Songwriters explore themselves and the world through this combination process, allowing to express their inner selves in ways that would not exist otherwise. It is

[24] https://musicforgalway.ie/cellissimo-2024/, accessed on 21 August 2024.

interesting to notice that, so far, no large attention has been given to the sociological analysis of the relationship between music and emotions, nor the study of the emotional connections with the ocean (Cobb, 2016; McKinley et al., 2023). Therefore, the experience of these music projects for the ocean that have happened in Galway since the end of 2020 offers a novel case study from both perspectives. In fact, the lively music scene in Galway, and Ireland in general, offers many opportunities to keep learning how music can help us explore our emoceans.

An important element to consider bringing forward is the participatory aspect of the songwriting process, which was present in all projects described here. In fact, as explained by Lewis (2022), the experience of working with someone to put their personal story of connection with the ocean into a song can be transformative for both the songwriter and the person/people who shared the story. It actually goes beyond that. By sharing these personal insights through a song, people who listen can possibly resonate with it and even more, this transformative experience can be amplified if the song is sung together by multiple people gathered by the social bonding power of music (Savage et al., 2021) to "*sing all our cares away*"—to quote the Irish songwriter Damien Dempsey[25]—for the ocean.

References

Beech, H. F. (2015). Songwriting and transformation: The subjective experience of sharing self through song. *International Journal of Transpersonal Studies, 34*(1), 24.

Buchan, P. M., Evans, L. S., Barr, S., & Pieraccini, M. (2024). Thalassophilia and marine identity: Drivers of 'thick'marine citizenship. *Journal of Environmental Management, 352*, Article 120111.

Cobb, M. C. (2016). "When i feel a song in me": Exploring emotions through the creative songwriting process. In *Symbolic Interactionist Takes on Music* (pp. 61–79). Emerald Group Publishing Limited.

Cox, L., Muñiz Piniella, A., Pakszys, P., Zielinski, T. (Eds). 2021. CommOCEAN 2020 conference proceedings. CommOCEAN 2020—4th International Marine Science Communication Conference. Sopot, Poland, 1–2 December 2020. ISSN: 2736–5778 https://doi.org/10.5281/zenodo.4530741

[25] https://open.spotify.com/track/5g0diCzmmi0zfMEglJOJgH, accessed on 30 August 2024.

Foley, R. (2022). Swimming: Immersive encounters in the ocean. *The Routledge handbook of ocean space* (pp. 298–310).
Gibson, J. M. (2009). The birth of the blues: How physics underlies music. *Reports on Progress in Physics, 72*(7), Article 076001.
Kearney, W. (2024, April). *Learning by ear: Multimodal listening and the embodiment of irish traditional music and dance.* In British Forum for Ethnomusicology & International Council for Traditional Music Ireland Joint-Annual Conference, 4–7. (Unpublished).
Keegan, N. (2011). The linguistic turn at the turn of the tune: The language of 'contemporary ensemble'in Irish traditional music. *Ethnomusicol. Ireland, 1*, 37–48.
Kelly, C., Gainza, M., Dorran, D., & Coyle, E. (2010). *Locating tune changes and providing a semantic labelling of sets of Irish traditional tunes.* International Society for Music Information Retrieval.
Koulouri, P., Mogias, A., Mokos, M., Cheimonopoulou, M., Realdon, G., Boubonari, T., Previati, M., Formoso, A. T., Kideys, A. E., Hassaan, M. A., & Patti, P. (2022). Ocean literacy across the Mediterranean Sea basin: Evaluating middle school students' knowledge, attitudes, and behaviour towards ocean sciences issues. *Mediterranean Marine Science, 23*(2), 289–301.
Lebreton, L., Slat, B., Ferrari, F., Sainte-Rose, B., Aitken, J., Marthouse, R., Hajbane, S., Cunsolo, S., Schwarz, A., Levivier, A., & Noble, K. (2018). Evidence that the Great Pacific Garbage Patch is rapidly accumulating plastic. *Scientific Reports, 8*(1), 1–15.
Lewis, M. (2022). Story-to-Song: Creating sustainability through songwriting. *Available at SSRN 4551149.*
Lotze, H. K., Guest, H., O'Leary, J., Tuda, A., & Wallace, D. (2018). Public perceptions of marine threats and protection from around the world. *Ocean & Coastal Management, 152*, 14–22.
McCullough, L. E. (1977). Style in traditional irish music. *Ethnomusicology*. 85–97.
McKinley, E. (2023, September). Ocean literacy and coastal connections in Ireland. *Final report produced for Fair Seas.*
McKinley, E., Burdon, D., & Shellock, R. J. (2023). The evolution of ocean literacy: A new framework for the United Nations Ocean Decade and beyond. *Marine Pollution Bulletin, 186*, Article 114467.
Ohs, C. L., DiMaggio, M. A., & Grabe, S. W. (2011). Species profile: Pigfish. *Orthopristis Chrysoptera.*
O'Shea, H. (2007). Getting to the heart of the music: Idealizing musical community and Irish traditional music sessions. *Journal of the Society for Musicology in Ireland*, 1–18.

Rapuano, D. L. (2001). Becoming Irish or becoming Irish music? Boundary construction in Irish music communities. *The Journal of American Culture, 24*(1–2), 103–113.

Savage, P. E., Loui, P., Tarr, B., Schachner, A., Glowacki, L., Mithen, S., & Fitch, W. T. (2021). Music as a coevolved system for social bonding. *Behavioral and Brain Sciences, 44*, Article e59.

Tranquada, J., & King, J. (2012). *The 'Ukulele: A history*. University of Hawaii Press.

Verne, J. (1916). *Twenty thousand leagues under the sea*. C. Scribner's Sons.

Whitworth, V. (2018). *Swimming with seals*. Bloomsbury USA.

Wootton, N., Nursey-Bray, M., Holland, S., & Gillanders, B. M. (2024). Better understanding ocean awareness: Insights from young people. *Marine Policy, 164*, Article 106159.

Zoom Video Communications Inc. (2016). *Security guide*. Zoom video communications Inc. https://d24cgw3uvb9a9h.cloudfront.net/static/81625/doc/Zoom-Security-White-Paper.pdf

CHAPTER 10

Behind the Scenes of a Collaborative Process to Make Blue Kinships Visible

Ruth Brennan and Stephen Hurrel

10.1 Setting the Scene

We live in a world of complex global environment-society challenges, from climate change, biodiversity loss and environmental degradation, to natural resource management, and food and energy security. It is a matter of societal choice how the world we live in ought to be. As such, policy and dominant societal narratives are underpinned by normative assumptions, specific worldviews and value systems that may conflict with values held by those who have not shaped these narratives. The arts, humanities and social sciences have a crucial role to play in providing deeper insights into human motivations, values, worldviews, and into the societal choices made to address these challenges (Holm et al., 2015). Art-science collaborations can challenge accepted ways of framing the

R. Brennan (✉)
Trinity College Dublin, Dublin, Ireland
e-mail: ruth.brennan@tcd.ie

S. Hurrel
Hurrel Visual Arts, Newcastle, UK

C. Certomà (ed.), *Blue Kinships*,
https://doi.org/10.1007/978-3-031-78619-8_10

people, objects and concepts that artists and scientists engage with by opening up different ways of looking at scientific issues and related societal challenges (Brennan, 2018a; Kuchner, 2022; Paterson et al., 2020). Transdisciplinarity—collaborative research with people and institutions outside academia—has significant potential for socially just societal transformation towards sustainability, not least within the blue sphere (Franke et al., 2022; Heidkamp et al., 2021).

Likewise, the expanded field of art encompasses a broad range of approaches and environments in which art activities can unfold. These include forms of collaborative art, participatory art and socially engaged art practice, creating opportunities to integrate and even camouflage art into real-world situations and contexts. Relationships between people, places, and things are foregrounded, rather than aesthetic qualities or artwork value. Critiques of socially engaged art practices have been ongoing, particularly since the 1990s (see Bourriaud, 2002 and Bishop, 2004). Bishop digs deep into the ethical implications of relational aesthetics and offers a counter narrative that not only recognises the shift from aesthetic objects to a more complex set of conditions that include relationships between artist-spectator, producer–consumer, and the wider social-political context, but also expands on the micro-utopia artworks cited by Bourriaud to include conflict and negative relationships, challenging viewers to question their own ethical positions in relation to the art they are confronted with. Chantal Mouffe advocates "a pluralistic approach, according to which the critical potential of art is not constrained to strictly negative and reactionary responses, but also bears the responsibility to put forward new models of politics and new modes of collective identity" (Miller, 2016, p. 175).

Transdisciplinarity and socially engaged art practices have formed the bedrock of our art-science collaborations that explore blue kinships.

10.2 Blue Kinships as a Contested Space in Our Art-Science Collaborations

Just as our oceans and coasts are highly politicised (Bennett, 2019), the concept of blue kinships can help interrogate contexts that may be presented as objective and value-free. Art can be deliberately provocative to facilitate such interrogation, in order to challenge accepted framings of issues. However, our art-science collaborations tended to be situated in already fraught and contested socio-cultural and political contexts.

For example, during the development of our Sea Stories project on the island of Barra in the Outer Hebrides, Scotland we found ourselves within a contested space that required us to tread carefully and listen closely in order to find appropriate forms of engagement that helped build trust between us and members of this small island community. The Sea Stories collaboration emerged from the context of an entrenched and longstanding conservation dispute between the Scottish Government and inhabitants of Barra. Since 2000, many of the islanders had fiercely resisted the creation of two European-driven marine special areas of conservation (mSACs) as part of the EU's Natura 2000 network, articulating a fear of losing control over their marine resources.

We were initially perceived as outsiders by the islanders. They were unsure as to which side of the conflict we were on, assuming that we had to be on one side or the other. Indeed, Brennan later discovered, that some islanders wondered whether she was an informant, planted in the community by the Scottish Government. This level of suspicion that we encountered at the outset returned to haunt us at various stages throughout our process. The fraught context on Barra required us to be keenly aware of the conflict in terms of personal and political dynamics within the community as well as any historical issues that may have nurtured a sense of mistrust. During the autumn of 2011, while undertaking research for her PhD, Brennan had started to build relationships through interviewing several key people involved in the conflict. As an engaged researcher, she was grappling with how to explain and distinguish her role as a researcher from that of an advocate for a particular side in the conflict, in order to maintain her credibility as an independent scholar. This is not an easy task for the engaged researcher, as the "hands-on and applied approach, which is implicit to engaged research, can often be interpreted as advocacy" and this, in turn, can lead to concerns about compromising the critical edge of the research being undertaken (Brennan, 2021, p. 312). Brennan was also searching for an angle to explore the conflict that was unexpected and not polarising, in order to provoke a more nuanced dialogue and ideas on potential ways forward.

Social scientist Iain MacKinnon was also on Barra during this time, conducting related research into the mSAC conflict. When Hurrel joined Brennan and MacKinnon on the island, initially to research and develop a short film he had been commissioned to produce, it became clear to all three that there were shared interests and approaches to engagement that could make it mutually beneficial to collaborate with each other. Significantly, just before Hurrel arrived on the island, Brennan and MacKinnon had attended a community consultation meeting about the proposed mSAC designation, hosted by Scottish Natural Heritage (the Scottish Government's conservation agency who were managing the mSAC designation process) and attended by about 70 islanders, many of whom were angry at the proposition of the designation of an mSAC in island waters. At that meeting, Brennan and MacKinnon noticed that the conservation map presented to the local community by the Scottish Government did not reflect the blue kinships between island inhabitants and their local marine environment, as the two-dimensional and static government map only reflected the biological diversity deemed worthy of special protection in the island waters. The interrelated socio-cultural relationships between islanders and their marine environment did not feature on the map. This led Brennan and MacKinnon to discuss the idea of a more dynamic map, to reflect intergenerational knowledge, fishermen's ways of knowing the sea and the intangible cultural heritage of this biodiverse and biocultural marine environment. Hurrel's arrival coincided with the beginnings of these conversations around a dynamic map of the sea and a boat trip with an island fisherman (Picture 10.1). As they were passing through a particularly rough stretch of water, the fisherman happened to mention that these waters were called "Caolas Boga' Mara"—the "Narrows of the Dipping Sea". Brennan asked if there were other names and stories in the island waters and he confirmed that they were there in abundance. On the same trip, the fisherman showed us a digital map of where he deployed his creels to catch lobster and crab. This inspired Hurrel to propose the idea of an interactive digital map which could convey the intangible cultural heritage that was intertwined with the rich biological diversity of the island waters.

During our collaborations on Barra, it became evident that our practices were characterised by a shared curiosity around different people's perceptions of the same landscape as a way to reveal hidden relationships within natural and human-made environments. In Clyde Reflections, a subsequent Hurrel and Brennan collaboration, the contested space of blue kinships was less evident than the conflict on the island of Barra. Creative Scotland commissioned us, along with 13 other art-science projects to "explore the interplay between the natural world and its representation, and promote deep collaboration and knowledge exchange between the creative and scientific sectors".

Picture 10.1 Gathering stories on a fishing boat off the coast of Barra, Outer Hebrides, Scotland (photo Stephen Hurrel)

In our introduction to the Clyde Reflections film, we observed that "[w]e see this project as a response to an environment that is often seen as picturesque, natural and seemingly unaffected by interactions between people and place. On the surface, from a distance, the Firth of Clyde looks fairly pristine, but it is only when you begin to understand the complexity of the environment, and begin to see the precarious balancing act that is going on—often beneath sea level—that your perception of it changes" (Hurrel & Brennan, 2014a).

We used the meditative, cinematic medium of Clyde Reflections to explore "the richness and diversity of (sometimes conflicting) perceptions of the marine environment of the Firth of Clyde in Scotland. Diving into the depths of this watery environment… we uncovered a marine environment which has been both altered by people and enriched by its intangible cultural heritage" (Hurrel & Brennan, 2014a).

This chapter explores our collaborative process to make blue kinships visible under three themes: methodology, ethics and vulnerability, and funding. We discuss how the overlaps in our methods and methodologies facilitated an enmeshed co-creative relationship that is distinctive from the usually more separate practices of artists and scientists in art-science collaborations (see for example Griffith, 2014). We delve into some of the ethical issues we faced during our collaborations with local communities and how our ethical alignment was invaluable in helping us to navigate this terrain. We provide behind the scenes insights into the crucial ingredient of vulnerability throughout our work and the sensitivity that was required to work with and within this emotional space. Finally, we consider how the flexibility of the funding support for our initial collaborations was instrumental in allowing the intertwining of our respective practices to flourish.

10.3 Methodology and Methods

For the initial collaborative project on Barra (the illustrated publication Belonging to the Sea (MacKinnon & Brennan, 2012)), Hurrel, Brennan and MacKinnon combined to create a group that was greater than the sum of its parts, providing a diverse yet complementary set of skills and experience. However, it was through the process of discussing ideas, developing engagement methods and working together that we recognised similarities in our approaches and how art and creativity may contribute to these synergies. Hurrel had developed engagement approaches through working collaboratively with other artists and specialists and through his work in a range of community and public settings to produce context-based artworks (see for example Hurrel, 2014 and Hurrel, 2019). In Brennan's engaged research approach, she combined established academic research methods with more experimental approaches to explore how artistic methods may lead to more engaging, creative outcomes (see Brennan, 2012, 2019).

In our first two art-science collaborations—the Belonging to the Sea illustrated publication and the Sea Stories online interactive map of the sea (Hurrel & Brennan, 2013)—

we wove together several rich veins to inform our collaborative process, including Brennan's PhD research and Hurrel's individual film work on Barra, all of which drew on and informed each other. This provided us with fertile ground for our art-science methodology (Picture 10.2) and methods to emerge intuitively and organically, and in a way that was sensitive to the fraught socio-cultural and socio-political context in which our collaboration was immersed. We conceptualised our methodology in hindsight, while reflecting on our process in Sea Stories in particular and driven by Brennan's need to articulate an art-science methodology in her doctoral thesis. Our methodology and methods were emergent and informed by our unfolding co-creative relationship. An important characteristic of our co-creative process involved having the creative freedom to work with a sense of the unknown and an ability to be vulnerable in process.

Key aspects of our process are to listen and remain open to the direction suggested by the material. Thereafter, we reflect on that material and try to navigate a path through it, in order to find the best format to communicate emerging ideas. For example, each of the projects we have collaborated on were initially either intended to be in a different format or the format was developed in response to the material, emergent ideas or changes in circumstances.

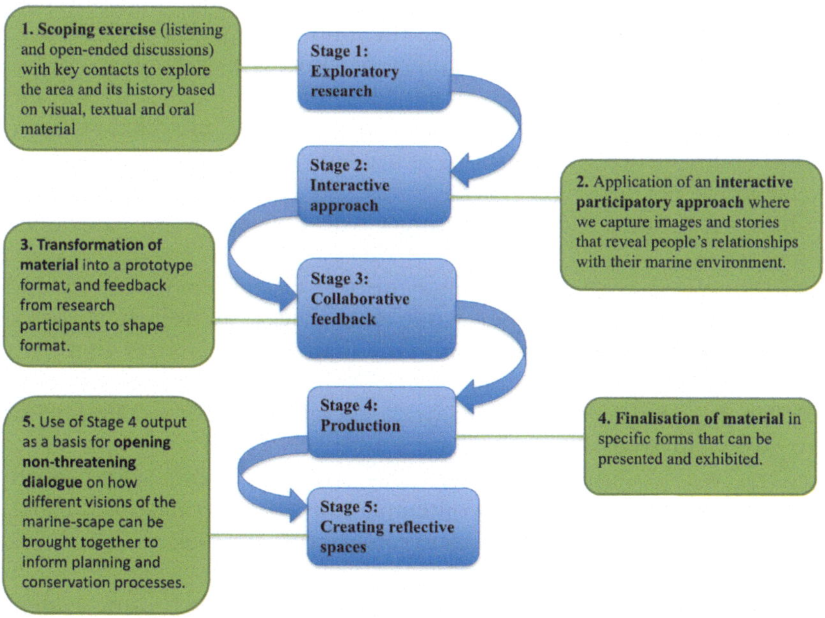

Picture 10.2 "Brennan and Hurrel art-science methodology" (Brennan, 2018a)

In Clyde Reflections (Picture 10.3), due to funding and time constraints, our approach to engagement had to become more structured. However, we also wanted a structure that was flexible enough to allow for intuition and happenstance. In particular, even though we adopted a documentary-style approach to explore participants' perceptions of the sea we were also interested in their individual working environments; the materials, equipment and types of visual imagery they would use, or create, in their daily working lives to help them navigate their understanding of the sea. As we had decided that only the voice of each participant would feature in the film, any objects or images relating to each person had potential to become a visual signifier of each participant and their particular relationship to the sea. This included microscopic photographs, videos of diatoms and cell structures, topological models of the seabed, hand-drawn diagrams of tidal flows, as well as archival photos. The most significant "find" was access to hours of underwater video footage filmed in Lamlash Bay on the island of Arran by conservationist Howard Wood. This video material became the foundation on which to build the film and to create the immersive quality that we were aiming for. In a similar way, the interview process that Brennan undertook for Clyde Reflections was designed to draw out the character and perspective of each individual but also to be flexible enough to deviate from the semi-structured questions as needed. This resulted in a rich and diverse range of voices.

On articulating our process during Clyde Reflections, we concluded that:

"Although we belong to seemingly distinct disciplines, we have found that the overlaps between our working practices, as well as a shared aesthetic sensibility, are more important than the differences that exist within our respective fields.

A defining characteristic of our work, therefore, is that we are able to draw on a range of different material and to respond to it in a structured as well as intuitive manner. This approach allows us to create work that opens up space for contemplation by reflecting the unfixed, shifting nature of relationships between people and place" (Hurrel & Brennan, 2014a, p. 49).

The emergent nature of our collaborative process also became evident in our approach to various ethical dilemmas. In the next section, we reflect on how it was serendipitous that we were aligned in our approach to navigating these dilemmas.

Picture 10.3 Clyde Reflections film (Hurrel & Brennan, 2014b) installed in Gallery of Modern Art, Glasgow, Scotland (photo Alan Dimmick)

10.4 Ethics and Vulnerability

Although we were perceived as outsiders by the islanders, this perception was tempered as time progressed, as each team member had a connection to Barra or Celtic/Gaelic culture. MacKinnon was a Scottish Gaelic speaker. Brennan spoke Irish, understood some Scottish Gaelic and shared heritage traditions common to Ireland and Scotland such as those around St Bridget's Eve/Day on 31 January/1 February. Hurrel had family connections to Barra through his mother (née MacNeil, the main clan of Barra), who was born and raised there. His grandfather had been a respected member of the island community through his crofting and fishing work and his service as lifeboat coxswain for 14 years, and was still remembered fondly by many of the older fishermen of Barra. This, however, did not shield us from difficult situations we faced as "outsiders".

Dealing with ethical dilemmas became a feature of our art-science-community collaborative process almost from the outset. This is perhaps unsurprising, since working to reveal blue kinships inevitably involves engaging with people's emotional connections to place. This concept of intertwined human-nature relationships is beautifully conveyed by the Irish and Scottish Gaelic words dúchas/dùthchas, which the Scottish Gaelic scholar John MacInnes describes as an emotional energy. “What is to a stranger an expanse of empty countryside—magnificent or drab according to prevailing notions—to the native sensibility can be dynamic, perhaps even heroic, territory peopled with figures from history and legend” (Newton, 2006, p. 29). While “dúchas/dùthchas” is a word of the land, in the illustrated publication Belonging to the Sea, MacKinnon and Brennan (2012) argued that the emotional energy of belonging to place and the responsibility the word conveys extends to the waters around the homeland: “dúchas/dùthchas na mara”—the sense of belonging to the sea. We soon discovered that navigating the watery terrain of the emotions can lead to stormy seas.

Several interrelated incidents on the island of Barra brought us face to face with ethical issues and the vulnerability inherent in working with a marginalised population. The island’s social, historical and cultural context over time contributed to the islanders sense of being oppressed, from forced migration due to extreme poverty and instances of mass evictions from unprofitable farmland with extortionate rents (within the wider context of the nineteenth-century Highland Clearances), to various terrestrial nature conservation designations in more recent decades, which many islanders felt were imposed without proper consultation. This victim-oppressor mentality has carried through to present day Barra and influences local perceptions of outsider involvement in local life (Brennan, 2018a, 2018b). Despite our awareness of the sensitivity of this context, we became caught up in its dynamics on several occasions.

The first issue arose when we offered to run an interviewing and video camera skills workshop with pupils in the local school, to encourage them to collect sea stories from family members for the recently funded Sea Stories project. The school was not comfortable with their pupils being involved as collectors in the Sea Stories project, because Brennan was carrying out research on the mSAC conflict. Two months previously, members of a local group called SHAMED (Southern Hebrides Against Marine Environmental Designations) had run a workshop with school pupils in creating a large banner which SHAMED subsequently displayed outside the Scottish Parliament in protest against the mSAC. The banner outside the Parliament, which later appeared in Scottish media, displayed the outline of the island of Barra containing the words “HANDS OFF OUR FUTURE. NO MSAC”, and was surrounded by multicoloured hand prints of the schoolchildren from the SHAMED workshop. The children's parents were upset that the school had not sought parental permission for their children to take part in the workshop and outraged that their children had been used by SHAMED, in their view, as political tools in the conflict. Understandably, both school and parents were now very sensitive to the schoolchildren being involved in any new creative project based on the sea. Brennan carefully explained that her role as a researcher in the project was not to support either side of the conflict and that the Sea Stories project was about reflecting the intangible cultural heritage in the seas around Barra. We also suggested that the school could ask parents' permission for their children to attend the workshop. By the end of the meeting, the school principal and media teacher appeared to be reassured and agreed that the workshop could go ahead with second year pupils (13–14 years old) as part of their media module.

Two days later we were taken by surprise when we attended a meeting with the Board of our future collaborators, Voluntary Action Barra and Vatersay (VABV). We had been invited to the meeting to make a presentation about the project and to show a prototype version of the online digital map. Certain members of the community organisation believed that they could take ownership of the project, including appointing their own technical team to produce it, and potentially commercialise the format to sell on to other island communities. In response, we pointed out the ethics of the free exchange of outputs of projects that are paid for by public funding and reiterated the intention of the project as being to co-create a Barra-specific map that would be owned, and continue to be populated, by the community via VABV. The map would record and safeguard the island's intangible cultural heritage for generations to come, as opposed to a commercialised product. We were keenly aware, however, that underlying this episode was a history of the islanders being marginalised and dispossessed. There was therefore an understandable need to ensure that they did not feel they were losing control of their cultural heritage, in the same way that the mSAC conflict embodied a fear of loss of control over their natural resources through restrictions on activities that a conservation designation might entail.

Working within an arts and community setting and engaging with people creatively can be a complex process when it comes to intellectual property and ownership. There have been many critical discussions about ownership of artworks when produced through a collaborative process. However, if there is clarity and agreement about the process and outcome, this should mitigate against misunderstandings later on different levels of collaboration can range from participation with an artist on process or outcome (which is usually an artist's concept, offered as an opportunity for collaborator participation) to co-creation which blurs the line between artists and collaborators. In co-creation an artist may work more as a creative facilitator, for example, with a starting point based around generating ideas, concepts and frameworks on which to develop a project in response to the interests of a group or community and their context. This may be partly led by the artist/facilitator and partly led by members of a community group. In this case, ethical issues need to be considered from the beginning of the project; about who has control and/or ownership of the project concept or physical outcomes as well as who is credited with the production of the artwork, event, etc. With any collaborative relationship there is always the possibility of imbalance; of shifts or exertions of power, however subtle these may be. It is particularly important for participants and organisers to be aware of potential ethical issues of working with people who are already in a vulnerable situation and feel they have no agency in their lives. We had erroneously believed that it was clear that Sea Stories was a collaborative project, based on a concept design by Hurrel and Brennan, which would provide a framework for a diverse range of individual and collective expression. As many creative projects evolve over time, the inevitable fluidity within the process and enthusiasm for the idea perhaps makes it more difficult to step back and discuss potential ethical issues, whether obvious or, in our case, surprising.

Together with concerns from the school over the workshop, it became increasingly evident that we were operating in a highly fraught and sensitive socio-cultural and socio-political environment and that any mis-steps on our part could potentially be fatal to the success of the Sea Stories project. As such, we were extremely careful to check with the community organisation, VABV, where we should launch the Sea Stories project and distribute printed copies of the related illustrated publication, Belonging to the Sea. We tried to integrate our events with pre-existing community events, where possible, so that this would be less demanding on people's time. On the advice of VABV, we launched Sea Stories at a ceilidh (an informal event featuring traditional music, songs, and dance) held for the Clan MacNeil Gathering, where MacNeils from the Scottish diaspora travel from America and Canada to the island of Barra for a bi-annual summertime clan gathering. We were puzzled as to why very few islanders attended.

Several months later, the reasons for this poor attendance became apparent. As well as a certain amount of disquiet from the school around the Sea Stories workshop, a member of the Heritage Centre (who had previously been very supportive of the project) challenged the Chair of VABV as to whether the Sea Stories project was beneficial to the community and expressed concerns around cultural heritage material leaving the island, with no benefit for the community. This turn of events was extremely worrying for both Brennan and Hurrel, given the attention we had paid to these sensitivities from the outset. It was clear that we had, however unintentionally, done something that led to feelings of exclusion amongst the very people we wanted to work with and who, we thought, understood the value of the project best. It transpired that the launch of Sea Stories at a Clan MacNeil Gathering event had not been a wise, or sensitive choice, as we had not realised that there was tension between local people and the organisers of the Clan Gathering. Traditionally, the Clan MacNeil Gathering had been a community-organised event that took place every 5–10 years. More recently, the event took place every 2 years, and was organised by an enterprising islander who spotted an opportunity to bring visitors to the island. This led some islanders to view the Gathering as an elitist event for a group of wealthy Americans and Canadians who could afford to travel to Scotland every two years. This feeling that the event was not for islanders was compounded by disagreements between the local drama group, who would normally be involved in the clan event, and the event organiser. In summary, the Heritage Centre member informed us that we had launched Sea Stories during the busiest week of the year and chosen a partner event that local people were actively avoiding. It was therefore perceived that Sea Stories was not being developed for, or with, them.

Relationships, once damaged, are difficult to repair in such sensitive and vulnerable contexts. Brennan thought she was able to repair, to some extent, the damaged relationships by explaining that Sea Stories had just started and that we had hoped that the Heritage Centre would be central to the project, with its focus on intangible cultural heritage. It also became clear that the Chair of VABV, who was a new appointment, had not yet built up the regular communication channels that the Heritage Centre had enjoyed with the previous Chair, which had contributed to this unfortunate misunderstanding. These events also revealed our own vulnerability to failure. As we employ an open approach to research and engagement, we do not necessarily know where that may lead in terms of success or failure of the process and/or final outcome. Additionally, as "outsiders" working within a community we often have to rely on others within the community to help us steer a smooth course.

Having discussed these issues in depth we (Hurrel and Brennan) were fortunate in that we found ourselves to be aligned in our approach. Neither of us wanted to continue a project that might be perceived as culturally dispossessing an already marginalised population. We agreed that the islanders needed to make this decision. Therefore, via the VABV Chair, they were asked whether they thought it best to close the project and return the funding to Creative Scotland. We were assured that they did not want to pursue this course of action. This allowed us to progress in the knowledge any conflicts or issues, whether justified or not, had been put aside and there would now be local support for the project. This alignment in our ethical approach was fortuitous as it was not something we had explicitly discussed beforehand. In our case, it strengthened our collaborative relationship and contributed to us wanting to continue working together in the future.

When the Sea Stories project officially ended after a year, we held a closing event in the Heritage Centre where we formally handed over the map to the islanders, into the care of the Heritage Centre. This event was well-attended by local people, which we took as a sign of relationships being re-balanced and of perceptions being positively reconfigured. At the event, Brennan's photo essay (Brennan, 2012, 2019) was shown as a slideshow with music, followed by examples of sea stories on the Sea Stories map, which featured the voices of some members of the audience. This provoked much discussion and laughter about other, unrecorded, sea stories and prompted one local to remark to Brennan, "you're very close, you've got underneath my guard" as they admitted to the vulnerability that the Sea Stories map (Picture 10.4) elicited. This telling comment points to another important feature of work that engages with emotions—it brings people (participants and project team alike) into spaces of vulnerability, where they may not feel entirely comfortable.

In our experience of working with fishermen on Barra we found them to be open, interested and engaged in the project. This may have been because we were taking an interest in, and valuing, something that was familiar to them but which perhaps they did not regard as being of particular interest to outsiders. In this kind of relationship, and exchange of knowledge, there is a certain amount of trust involved and therefore it is important to approach it with sensitivity. An example of this would be

Picture 10.4 Sea Stories: Barra—an online cultural map of the sea (Hurrel & Brennan, 2013)

when local fishermen shared their individual fishing spots and revealed the triangulation method they would use to find them while at sea. We were concerned that they might not want to make this publicly visible. However we were assured that we could include this on the map as the fishing spots were localised areas and not likely to be over-used by visitors to the island.

In terms of the actual interview process there were certain considerations taken into account that we felt would help to close the gap between interviewer and interviewee. This included conducting interviews in locations where participants felt most comfortable, such as their homes or on their boats. Another aspect that Hurrel felt was important in recording the interviews, was to be as unobtrusive as possible in terms of equipment set-ups, as a way to achieve a relaxed environment that was conducive to conversation. Therefore, any first-time interviews used domestic scale video and audio equipment as these were more informal, and less intimidating than a large-scale camera on a tripod, with an oversized microphone. This resulted in a more conversational approach with a focus on the dialogue, stories and relationships between people in the room, without too much distraction from the recording process.

10.5 Conclusions

In this chapter, we have reflected on our collaborative art-science process over the course of three different collaborations. As it is unusual to read about the day-to-day trials of such collaborations, we have aimed to share, in particular, our insights into the challenges we traversed while working in sensitive and politically fraught community contexts. We have also shared insights into the participatory and creative practices we used to engage with small island and coastal communities in articulating their emotional attachment to place, and how we navigated through the engagement and development process. By reflecting knowledge and insights back to participants in the format of a film, photographic installation, publication or online digital artwork a new "space" is created where ideas can meet, converge, and become visible and tangible.

Having the time and space to experiment has been crucial to our process which is characterised by the unplanned and unforeseen, allowing space for the unexpected to emerge. Facilitating a truly responsive engagement process—one that is aware of, and willing to work within, potentially complex and conflicting relationships within a community or context—requires recognition from institutional decision-makers in the policy, academic, and arts worlds of the value of this type of transdisciplinary activity. The creative approaches of the arts, humanities and social sciences can help to open up new spaces for dialogue by engaging with people in culturally meaningful and relevant ways and to understand the nuanced relationships between people and place, but need to be supported to do this sensitive, time-consuming, and often unpredictable work.

Funding Working creatively with funding limitations can often lead to interesting outcomes. In Belonging to the Sea, a main shift was to think about the output as being something tangible and aesthetic that a wide range of people could engage with, rather than an extensive written report. This led us to think about the publication as a creative design object in itself and, with the benefit of having a wide selection of video and photographic material from the interviews and field trips, we felt that ideas being explored through the texts could be enhanced by this material, in particular by providing visual insights into the activities and environments of the two islands. To enhance the material qualities of the publication itself, and readers' experience of it, Hurrel worked with a designer on the layout, typology and book cover design as well as exploring options for the quality of paper, all of which was discussed and agreed with the team and which added to the aesthetic and tactile experience of the book. In particular, Hurrel chose a colour for the book cover that was an exact match with one of the familiar turquoise colours of the sea around Barra, sourced from a photograph taken by Hurrel. The final outcome was a forty-eight page, full colour publication in English, Irish, and Scottish Gaelic. Many of these publications were given away to attendees at the Sea Stories launch event and other community events on the islands, thus ensuring a wider readership beyond academia.

Similarly, when the Sea Stories funding award was 50% of the original budget, Hurrel's and Brennan's fees were halved but retained the full fee for subcontracted professionals to ensure the desired outcome of a fully functioning online map. The project management aspect, by Hurrel Visual Arts, required at least twice as much time as was allocated in the reduced budget. As the project was a combined creative and technical endeavour, involving a certain amount of development work and flexibility in response to new material that was being gathered, a significant amount of contingency was needed, but was not available within the reduced budget. In order to complete the project to the high standard we envisaged, unpaid additional project management time was provided by Hurrel Visual Arts. In addition, the computer programmer, who had previously worked on several projects with Hurrel, generously stretched his time to meet the needs of the final project outcome, as did Brennan, whose additional time and costs were covered by funding support from the Marine Alliance for Science and Technology for Scotland via the Scottish Funding Council. The unusually open nature of Brennan's funding, which supported her to develop a new area of social science, provided her with the flexibility to devote this additional time developing the Sea Stories art-science collaboration.

The reduced budget for Sea Stories also impacted the hand-over process in terms of embedding it within the community. As with many small community organisations, VABV's resources and activities were stretched to the limit allowing no additional time to be allocated to Sea Stories after the project ended. The reduced budget had no contingency to extend the project beyond the hand-over stage. An ideal scenario would have been to fund an enthusiastic local person to continue to promote the project and engage with islanders in gathering stories and photographs to add to the map. Although we achieved our aim of delivering a co-created online cultural map of the sea, working with funding that does not match the ambitions of a project can be frustrating for all involved. Nevertheless, Sea Stories has continued to exist online since it was launched in 2013 with VABV continuing to fund the website hosting fees up to 2022. Thereafter, Hurrel's computer programmer generously agreed to host the website without cost on his personal server. To date, Sea Stories continues to receive positive reactions from Barra residents, other island communities, as well as from many others in the arts and academic communities internationally.

References

Bennett, N. J. (2019). In political seas: Engaging with political ecology in the ocean and coastal environment. *Coastal Management, 47*(1), 1–21.

Bourriaud, N. (2002). *Relational aesthetics*. Presses du reel.

Bishop, C. (2004). Antagonism and relational aesthetics. October, 110, 51–79.

Brennan, R. (2012). *What lies beneath: Probing the cultural depths of nature conservation conflict in the Outer Hebrides*. Photo essay. Ruth Brennan, Oban.

Brennan, R. (2018a). Re-storying marine conservation: Integrating art and science to explore and articulate ideas, visions and expressions of marine space. *Ocean & Coastal Management (Special Issue: Coastal Systems in Transition: Interconnecting the Social Dimensions in Response to Coastal Change), 162*(1), 110–126. Picture 2 and Picture 4 were first published in Brennan 2018a. Copyright Elsevier.

Brennan, R.E. (2018b). The conservation 'myths' we live by: reimagining human-nature relationships within the Scottish marine policy context. *AREA (Special Section: Cultural Geographies of Coastal Change: Place, Identity and Vulnerability), 50*(2), 1–10.

Brennan, R. (2019). *What lies beneath: Probing the cultural depths of nature conservation conflict in the Outer Hebrides. Audio-visual installation*, Art in the Anthropocene conference, Science Gallery, Trinity College Dublin, Ireland, 7–9 June 2019. Retrieved from [https://vimeo.com/555734658]

Brennan, R. (2021). Reflections on methodological tensions in doing qualitative research at the science-policy-community interface. In M. Gustavsson, C. White, J. Phillipson, & K. Ounanian (Eds.), *Researching People and the Sea: Methodologies and traditions* (pp. 309–321). Palgrave Macmillan.

Franke, A., Peters, K., Hinkel, J., Hornidge, A. K., Schlüter, A., Zielinski, O., Wiltshire, K. H., Jacob, U., Krause, G., & Hillebrand, H. (2022). Making the UN Ocean Decade work? The potential for, and challenges of, transdisciplinary research and real-world laboratories for building towards ocean solutions. *People and Nature, 5*(1), 21–33.

Griffith, D. (Ed.). (2014). *Imagining natural Scotland*. Creative Scotland.

Heidkamp, P. C., Garland, M., & Krak, L. (2021). *Enacting a Just and Sustainable Blue Economy through Transdisciplinary Action Research, 189*(2), 246–258.

Holm, P., Adamson, J., Huang, H., Kirdan, L., Kitch, S., McCalman, I., Ogude, J., Ronan, M., Scott, D., Thompson, K. O., Travis, C., & Wehner, K. (2015).

Humanities for the environment - A manifesto for research and action. *Humanities, 4*(4), 977–992.

Hurrel, S. (2014). http://www.hurrelvisualarts.com/wp-content/uploads/2014/07/Stephen-Hurrel-Selected-Work.pdf

Hurrel, S. (2019). http://www.hurrelvisualarts.com/wp-content/uploads/2023/02/0.-HURREL-PUBLIC-ART-Folio_Sept-2019_PDF_compressed.pdf

Hurrel S., & Brennan R. (2013). Sea Stories online cultural map of the sea. https://www.mappingthesea.net/barra/

Hurrel, S., & Brennan, R. (2014a). Clyde Reflections: A film and audio-visual installation. In D. Griffith (Ed.), *Imagining natural Scotland* (pp. 38–51). Creative Scotland, Edinburgh and Glasgow. Retrieved from [http://www.hurrelvisualarts.com/wp-content/uploads/2018/04/Hurrel-and-Brennan-2014_Clyde_Reflections.pdf]

Hurrel, S., & Brennan, R. (2014b). *Clyde Reflections: A film and audio-visual installation*. Retrieved from [http://www.vimeo.com/89793693]

Kuchner, U. (2022). The value of art science: Improving the balance in collaboration practices between artists and scientists can impact knowledge production. *Writing Visual Culture, 10*, 23–42.

MacKinnon, I., & Brennan, R. (2012). *Dùthchas na Mara/Dúchas na Mara/Belonging to the Sea.* Exploring the cultural roots of maritime conflict on Gaelic speaking islands in Scotland and Ireland. Retrieved from [https://www.mappingthesea.net/Belonging-to-the-Sea.pdf]

Miller, J. (2016). Activism versus antagonism: Socially engaged art from Bourriaud to bishop and beyond. *FIELD Journal of Socially Engaged Art Criticism, 3*, 165–183.

Newton, M. (Ed.). (2006). *Dùthchas nan Gàidheal: Selected essays of John MacInnes*. Birlinn.

Paterson, S. K., Le Tissier, M., Whyte, H., Robinson, L. B., Thielking, K., Ingram, M., & McCord, J. (2020). Examining the potential of art-science collaborations in the Anthropocene: A case study of catching a wave. *Frontiers in Marine Science, 7*, 1–13.

CHAPTER 11

Ocean Connections-Animating the Ugly Duckling Between Storytelling and Science. Developing a SciArt Project

Kristin Bergaust

11.1 Introduction: The Oslofjord: Ocean Connections as an Artwork

The background for Ocean Connections, an artistic work based on the Oslofjord, is both the individual experience of the fjord as a natural habitat as well as the recognition of how the fjord shapes and influences the daily lives of its population. Data from scientific monitoring are also important prerequisites for the work." The Oslofjord is nature, economy, infrastructure, biology, materiality, culture, and history as well as future plans and challenges. A fjord is a geographical freak, a topography that contains and harvests the ocean and forces the sea to coexist with the inland, the agricultural, the industrial, the cultural, to become infrastructure and enter human perception as a spectacle" (Bergaust, 2020, p. 12).

K. Bergaust (✉)
Oslo Metropolitan University, Oslo, Norway
e-mail: bergaustkristin@gmail.com

C. Certomà (ed.), *Blue Kinships*,
https://doi.org/10.1007/978-3-031-78619-8_11

IMAL Art Center for Digital Cultures & Technology, Brussels 2024: Two black walls forming a corner are filled with projections of animated graphic images. A recurring figure reminiscent of a bird is visible to the left, on the right wall you can often see details of the larger figure or other images of details that are not visible to the human eye. Keywords and notes appear in white handwriting, as if on a blackboard. Directional loudspeakers allow you to listen to a digital soundscape. (Picture 11.1).

This artwork was the result of Ocean Connections, an interdisciplinary artistic research project by Kristin Bergaust as the lead artist in collaboration with scientists Guillermo García Sanchéz CTO Digital Earth Solutions Ph.D. in Applied Mathematics Madrid and Evangelos Voukouvalas, External consultant at the DG Joint Research Centre, Italy. Guillermo contributed his mathematical modelling of ocean pathways developed with Lagrangian Descriptor, while Evangelos helped finding, depicting, and interpreting relevant data material. The collaboration was set up in the context of NaturArchy by the SciArt project at European Commission's JRC Joint Research Center in Ispra, Italy. In the following, the different aspects of the work and how it was conceived is described and analyzed.

Picture 11.1 Ocean Connections video installation in IMAL, Brussels 2024 (Photo Kristin Bergaust)

11.2 The Interdisciplinary Team Work

In creating the artwork Ocean Connections, we were an interdisciplinary team working together over time. Guillermo, the expert on the Lagrangian Descriptors modelling, and I met in Ispra at the NaturArchy summer school in 2022. We did a field trip together in the Oslofjord in October the same year to formulate an idea for the NaturArchy. When our concept was accepted for further development, we were invited for a one-week residency in the JRC at Ispra to meet and discuss with scientists and researchers. This is when we first met Evangelos Voukouvalas, researcher in the JRC. As the project was developed, we met regularly online for about 18 months, until May 2024 when the works was mounted and presented in Brussels. In June 2023 we spent a five day residency in JRC Ispra for a final residency before developing the work for the exhibition version now on view in IMAL. In our online meetings we have reflected on and discussed oceanic ecological conditions both specifically and in general, shared scientific and artistic perspectives and as a result generated new ideas. Often new developments of the video animations were tested in these meetings, quite frequently I discovered new ideas or trajectories to research after our exchanges. Ocean Connections is now the resulting artwork: a two-channel video installation with sound that has been formulated and constructed through a dialogue between the scientists and myself for almost two years. Our process was supported by the SciArt project in JRC Ispra, especially the curators Caterina Benincasa and Claudia Schnugg who contributed their input on intermediate examples of the work. Project leader Adriaan Eeckels provided support of the overall concept and with budget and technical issues, while Virginia Bernardi took care of website, communication, and marketing. (SciART Project EC). I was also invited by Stephen Davies of DG Mare to show a work-in-progress version of Ocean Connections at the European Maritime days in Brest in May 2023. This was very useful as a test run, both technically and in terms of audience responses. The work was shown on a big empty wall in a conference centre, there was no context to mark the wall projection as art, except from there being an information stand. It was however noticeable that members of the audience that showed interest, tended to have professional research knowledge of ocean ecologies or connected subjects. All in all, the test run made me confident of the mode of display: a direct wall projection from the floor up. To test the size as more-than-human was also important to the continued development of the work, since the protagonist Ugly Duckling images were meant to be seen as larger than a human. In my experience, as a person in a built room, images you meet in an exhibition space has to be slightly larger than yourself to be perceived as the same size. To experience an image as larger than yourself, it has to be markedly so. Another take-away from Brest, was that it provided a certain encouragement that scientists working in the field seemed to find the work interesting and relevant. They also visually recognized the topics addressed in the work. The first iteration of Ocean Connections is quite similar to the left wall of the installation work. The theme is an investigation of biodiversity, how species develop in a certain geographical area. Fish species have varied greatly over the years, many species have become close to extinct. For instance, Thunnus Thynnus, the blue tuna was last found in 1962. In 2021 the cod is hard to find, accordingly fishing for cod is prohibited. The cod have been important to fisheries as well as private households, the video highlights some examples of species that have disappeared, and also shows some of the species that have entered these waters as ecological conditions have changed (Picture 11.2).

Picture 11.2 Still from Ocean Connections left wall. Thunnus thynnus (photo Kristin Bergaust)

11.3 The Oslofjord: A Social and Cultural Context of the Main Actor of Ocean Connections

The protagonist of the video installation Ocean Connections is the Oslofjord, inspired by different projections of the map and constructed from mathematical modelling of pathways in the sea. The Oslofjord runs from Skagerak, the ocean strait between Denmark and Norway. It is said to be a distance of 120 km from the inlet in the south to what we can call the head crowned by Norway's capital Oslo in the top northern end. A point of departure for my work with the Oslofjord since 2016, is that we cannot really tell where a fjord ends. Every process and activity in the fjord that affects biodiversity, such as eutrophication and pollution also affect the life of the global oceans to some extent. The content of the work follows closely what has happened in this specific fjord and how its local conditions have developed, the title Ocean Connections points to the fact that an ocean is never isolated, but connected to the global oceans. The Oslofjord was formed by volcanic activities into a rift valley approximately 300 million years ago. The valley was then filled with sea water. The Oslofjord is challenging to navigate, with shallow waters, many underwater rocks and formations that are not found in the famous, scenic fjords in the west of Norway which are all formed by glaciers filled with sea water. The ship traffic passing through the Oslofjord today is the most intense in any Norwegian port. Cruise ships, tankers and barges are frequent, small fishing vessels and all kinds of leisure boats fill the fjord (Bergaust, 2020).

In the west part of Norway, Sognefjord and Hardanger fjord being the most famous, scenic fjords attract tourists both over land and in cruise traffic. Tourism in the fjords in this part of Norway has a long and honourable history, being visited by the emperor William of Preussen, other German and British aristocracy and famous artists and writers for a hundred and fifty years (National Library, 2024). The western fjord landscapes are also prevalent in the national romantic era of Norwegian painting in the mid-nineteenth century, aimed at constructing a Norwegian identity based on an unspoilt farmer's cultural traditions. One much-loved example of this era is the famous painting Bridal Procession on the Hardangerfjord by Hans Gude and Adolph Tidemand from 1848 (Picture 11.3), in the collection of the Norwegian National Museum. Here the Hardangerfjord is a scenography to celebrate authentic popular culture to underpin the Norwegian independence built on the constitution from 1814, which again lead to the actual independence of Norway, breaking a union with Sweden in 1905.

The painting was completed in the studios of the Art Academy of Düsseldorf, Germany, based on the painter Hans Gude's sketches from the area of Hardangerfjord. Adolph Tidemand painted the participants in the wedding procession with much detail. This painting does not depict a specific geographical place, but is an idealized rendering closely connected to German romanticism of the beautiful fjord landscape to be admired. (University of Oslo).

The Oslofjord on the other hand, holds together the most populated areas in Norway, both west and east of the fjord are important areas for industry, agriculture, and forestry. Ocean Connections as a contemporary artwork, aims to tells stories from the fjord's perspective which is however closely related to human activities. Along the shores of the

Picture 11.3 Bridal Procession on the Hardangerfjord. Hans Gude and Adolph Tidemand 1848 (Pic. The National Museum /Børre Høstland)

fjord, 1,6 million inhabitants (a quarter of Norway's population) live in eleven towns and several smaller settlements. The most important city is Norway's capital, Oslo, with around 1 million inhabitants, counting the adjacent municipalities.

The Oslofjord is characterized by how its population make use of it. Pollution from sewers, agriculture, forestry and industry create immense problems. Nitrogen and phosphorus are at times too plentiful and stimulate growth of algae, leaving the fjord water with too little oxygen. At the same time many inhabitants around the fjord are strongly connected to a culture of leisure, being on and in the fjord in the summer, swimming, sailing, or paddling, spending holidays and weekends in cabins, on boats or camping by the fjord. Traditionally, women and children of the economic élite would live in a summer house from spring to autumn, with the husband and provider working in central Oslo, coming to visit on the weekend by small steamer services, wine served on board. Working class population would set up a camp for the summer in the islands and use a rowing boat to reach their families. Contemporary harbourfront developments has resulted in a very trendy public sauna and swimming activity along the fjord in the inner city of Oslo.

11.4 The Oslofjord as an Ugly Duckling Telling Stories with Lagrangian Descriptors

The comparison between the revered touristic west Norwegian fjords and the densely populated Oslofjord with the characteristic birdlike silhouette (Picture 11.4) reminds us of the literary fairy-tale The Ugly Duckling by Danish writer Hand Christian Andersen.

The tale relates the story of an ugly duckling born in a pond in the countryside. He is severely bullied by his siblings for being so strange and different from them. The duckling flees from his home and returns later as a young adult. Astonishingly, he has then become a beautiful swan, not an ordinary duck like the others (HC Andersen Senteret). This tale instilled the image of the Oslofjord with a nickname and a certain narrative energy. The Oslofjord has a story to tell of being less appreciated, of being different and in need of a visible identity to be proud of.

Natural sciences often lean on numeric data, at the outset of Ocean Connections however, our conception of the fjord resisted measurements to construct a geographical entity. This inspired a reflection upon how a

Picture 11.4 Still form Ocean Connections: The Ugly Duckling (Photo Kristin Bergaust)

natural environment might have different delimitations to that of a cartographic and political unit. However, the Lagrangian Descriptors models based on data from the Norwegian Meteorological Institute, were a solid numerical base of the modelling and also confirmed how different areas are connected by travelling organisms, pollutions or particles unknown. The modelling provided a visual three-dimensional depiction of a water body. The difference in colours describes pathways that cannot be crossed by algae, pollutions or oil spills, a system of inherent forceful movement in the ocean itself that is not visible on the surface, it cannot be seen by our plain eyesight. As the lead artist, I was responsible for the overall artistic concept and production, visualizing and editing elements of information and fantasy. The mathematical modelling of Lagrangian Descriptors by Guillermo García Sanchéz (Picture 11.5) was the vital factor to develop a concept and working process. Here is how the modelling can be explained: the Lagrangian descriptor measures the total distance travelled by a fluid particle starting from a specific location over a given time interval, both forward and backward in time. It essentially quantifies the complexity of the particle's path in the flow field. By calculating the Lagrangian descriptor for a grid of initial points in the flow field, one can visualize these LCS as sharp ridges or valleys in the resulting M field. This makes the Lagrangian descriptor a powerful tool for uncovering the hidden geometric structures that govern transport and mixing in complex, time-dependent flows like ocean currents (García Sanchéz, 2024 personal communication).

These models are highly beneficial as technological tools, as they can show where and when matters or particles have been carried, which can be calculated both in future and in the past. Following this, the technology computing these transports are very useful in many respects, for instance in finding and identifying harmful oil spills. As an artistic actor, determining much in the Ocean Connections, the Lagrangian Descriptors contributes something that adheres to the information of what it really is: it shows a strength and inherent force in a body of water that constitutes the Oslofjord, a visual sign system that implies something different than the surface imagery. Seeing the visual representations of

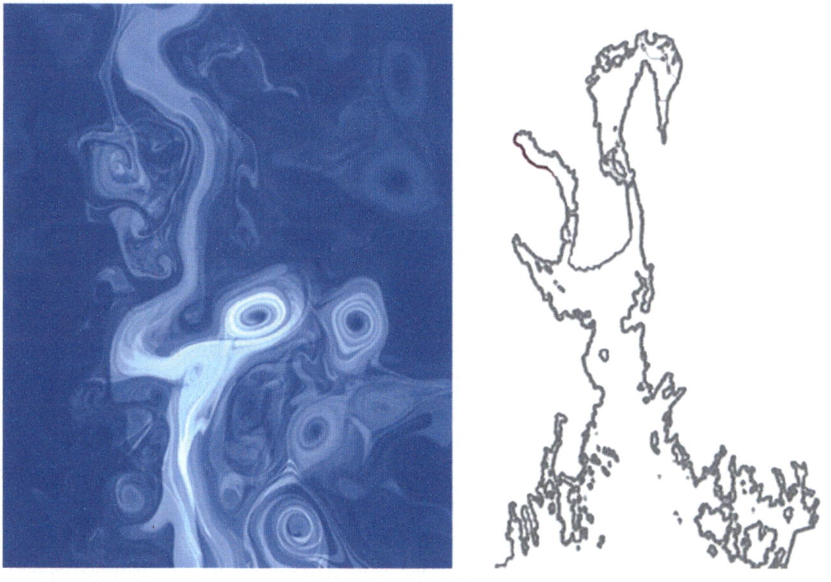

Picture 11.5 Left image made with Lagrangian descriptors, by A. M. Mancho, J. Curbelo, S. Wiggins, V.J. Garcia-Garrido, C. Mendoza: Beautiful Geometries Underlying Ocean Nonlinear Processes. Right image digital drawing by Kristin Bergaust. Oslofjord contour: The Ugly Duckling

this modelling when I met Guillermo at the Summer School of NaturArchy at the SciArt project in the JRC at Ispra in 2022, I was reminded of work I had done with three-dimensional water movements registered on video in my early years as an artist. I was taken back to Wasserwerk, my graduate project in collaboration with fellow student Nicola Schröder at the National Academy of the Arts in Oslo in 1991 where we used our body movements to create three-dimensional patterns though water bodies held by three-dimensional geometric Perspex containers. We knew about chaos theory, fractal geometry and fluids (Bergaust, no year).The possibility of modelling the Oslofjord with Lagrangian Descriptor as a tool opened new possibilities for storytelling, including my work with Oslofjord Ecologies since 2016 (Bergaust et al., 2020).

11.5 The Oslofjord in a Dire Ecological Crises

The Oslofjord is home to numerous organisms, it was once the most species-rich fjord in Norway. This abundant biodiversity is now severely endangered by the effects of human activities. Marine biologists tell us that these threats resemble what we find elsewhere in urban seas: insufficient cleansing of sewers and agricultural run-off cause eutrophication. The result is blooms of phytoplankton, causing lack of oxygen and poorer light conditions for marine organisms. One consequence of this is the growth of filamentous algae so called "lurv" which harm other organisms by overgrowing them. The lively "blue forest" is affected, causing a lack of for instance eelgrass meadows, which are needed by cod and other species as a protected living area for their young offspring. In addition, darker water reduces the conditions for photosynthesis and for filtering animals living on the seabed. We also see that some invasive species such as Pacific oysters can displace native species such as blue mussels and cause lack of nutrition for sea birds and other animals. Conditions in the Oslofjord show that urban development in the coastal zone and overfishing are examples of direct human interference with the ecological balance. However, these threatened species in the Oslofjord have not yet reached an ecological tipping point. This is where species disappear, not to return. It is still possible to restore conditions and re-establish plants and animals in the fjord. (Rinde, 2023).

11.6 The Oslofjord as a Site to Formulate Eco-Aesthetics

Trying to formulate a way of working with nature as an artist, I was searching for the Oslofjord as a subject, to let the fjord express itself and be present in a form not governed by the human gaze and at the same time engage human senses and emotions. Scientific tools and methods might provide exits and alternative viewpoints to escape the anthropocentric, stale, and sentimental renderings. Looking at a process rather than a photographed decisive moment changes the relationship to what it means to see an event. To experience a course of events through measuring and modelling instead of as a narrative of text, photo, or moving image is challenging. However, this might lead to perceptions of the environment that could denote a more-than-human viewpoint.

Western art history and visual culture has an established mode of representing nature from an anthropocentric viewpoint. Examples can be renaissance construction of perspective based on the human eyesight and symbolic renderings of natural phenomena as representations of human emotions, prevalent in romanticism and throughout modernism. Much as I admire these seminal artworks, I search for other perspectives. I am inspired by feminist film theory demonstrating the need for freeing the camera or visual apparatus from a certain ideology of representation, such as Laura Mulvey's influential essay Visual Pleasure and Narrative Cinema where she coins the term "the male gaze" as a phenomenon in Hollywood films (Mulvey, 1975). Although objectivation of the female body is by no means extinct in our time, contemporary culture is aware and sensitive to the effects of the male gaze in visual communication. We might need a parallel cultural shift in our representations of other species and of environments. Could questioning the human gaze create a more balanced representation of other living beings? By balanced, I think of a representation that gives credit to the other organism's characteristics, considering its senses and abilities. There are faculties and competences present in other beings that our senses cannot fathom. It might be that we could use our senses differently by new manners of stimulation or to enhance and change our experiences through technology. Visual and aural technology that do not mimic human senses and interpretations could be utilized (modelling, microscopy, radiology, medical imaging techniques, 360 photo and video among other possibilities). I also wanted to try out scientifically generated material with intuitively engaging aesthetic qualities such as the Lagrangian Descriptor modelling images. Appearing in the work are pathways in the ocean, not visible to the human eye. To search for artistic solutions and materials that can represent the Oslofjord in an emotionally engaging way motivates the creation of this work, it could suggest that the audience also belong to and are connected with the lifeforms and organisms that we normally cannot see.

Indigenous and agricultural traditions and knowledges might guide and inspire a less anthropocentric and more holistic aesthetics of the connections and possibilities in our way of perceiving the environment.

Experimenting with methods combining technology, ecology, and eco-aesthetic ideas might lead to new conceptual resources and enable imagination and fantasy to play along with scientific information. Donna Haraway's concept of sympoiesis meaning "making with" from her seminal book *Staying with the Trouble. Making Kin in the Chthulucene* solidifies and answers some of these ideas as the basis of methodology for the creation of an artwork. The concept of sympoiesis is stimulating, summed up by Haraway as "earthlings are never alone" (Haraway 2016, p. 58). She also adds this description of what sympoiesis might be: "perhaps as sensual molecular curiosity and definitely as insatiable hunger, irresistible attraction towards enfolding each other is the vital motor of living and dying on earth. Critters interpenetrate one another, loop around and through one another, eat each another, get indigestion, and partially digest and partially assimilate one another, and thereby establish sympoietic arrangements that are otherwise known as cells, organisms and ecological assemblages."

This biological drama and the description of the entanglements of everything living, emanates a freedom fed by fantasy and speculation as well as scientific facts. Haraway's ideas were engaging and productive in formulating a poetic for a work which at the outset had few formal prerequisites, except from the artistic repertoire of the artist. Trying to stand back from the human perspective, also makes it possible to investigate ecological imbalances and wicked problems without hitting the apocalyptic nerves so common in culture today. A more epistemological principle sourced from Haraway was also at work in discussing and thinking about Ocean Connections: a concept of situated knowledge which is here formulated as tentacular thinking: "It matters what thoughts think thoughts. It matters what knowledges know knowledges. It matters what relations relate relations. It matters what worlds world worlds. It matters what stories tell stories." This contributes to the belief I have in visual arts as a sensuous language that carries a lot of extraneous information. It matters whether the Disney Company, National Geographic or a marine biologist show you what a fish looks like. Life forms and environmental conditions coinhabit a complex, intertwined and largely unpredictable ecosystem, which also encompasses our lives. Humans are watery bodies; we are part of the water cycle, of the ocean and its flows. It's primordial waters, in which life was born, fill our cells. Whatever we do to the oceans, we do to ourselves. Consequently, we share water with other living beings. As Astrida Neimanis writes in her inspiring book *Bodies of Water*: "-the human is always also more-than-human. Our wateriness verifies this, both materially and conceptually." (Neimanis, 2017, p 5).

The search for an eco-aesthetic, is a shift in artistic and cultural language and imagination to make visible and present our entanglement with and interdependence of our living environments. Changes in sensuous input might influence human conduct, instil empathy and action and make us more understanding and communicative with non-humans and each other.

References

Bergaust, K (2020): Introduction: From experience to extension. In K. Bergaust, R. Smite, & D. Silina, (Eds.), *Oslofjord ecologies. Artistic research on environmental and social sustainability.*

Bergaust,K. (2024). https://www.bergaustvideo.net/wasserwerk-1991-93

García Sanchéz, G.(2024). E-mail to Kristin Bergaust June 3, 2024.

Haraway, D. (2016). *Staying with the Trouble. making Kin in the Chthulucene.* Duke University Press.

HC Andersen Senteret. (2024). The ugly duckling. Available at https://andersen.sdu.dk/vaerk/hersholt/TheUglyDuckling_e.html (Accessed 15th of June 2024)

Mancho, A. M., Wiggins, S., & Curbelo, J., et al. (2013). Lagrangian descriptors: A method for revealing phase space structures of general time dependent dynamical systems. *Communications in Nonlinear Science and Numerical Simulation.*

Mendoza, C., & Mancho, A. M. (2010). Hidden geometry of ocean flows. *Physical Review Letters.*

Mulvey, L. (1975). Visual pleasure and narrative cinema. *Screen, 16*(3), 6–18. Autumn 1975.

NASA https://earthobservatory.nasa.gov/images/46940/oslofjorden-norway

Neimanis, A. (2017). *Bodies of Water. Posthuman feminist phenomenology*. Bloomsbury.

The Intergovernmental Oceanographic Commission of UNESCO. (2024). What is Ocean Biodiversity? Available at https://oceanliteracy.unesco.org/ocean-biodiversity/ (Accessed 24.02.2024)

Rinde,E. (2023). Conversation with Kristin Bergaust March 6th 2023.

SciArtProject, European Commission. (2024). Resonances IV The fourth edition (2022–2024) of Resonances, the flagship initiative of JRC SciArt. Available at https://science-art-society.ec.europa.eu/resonances-iv (Accessed 15th of June 2024)

University of Oslo. (2024). Brudeferden i Hardanger. Available at https://wiki.uio.no/hf/ifikk/kun1000/index.php/Brudeferden_i_Hardanger (Accessed 15th of June 2024)

CHAPTER 12

Tidal Kinships: Sharing and Caring for Rhythmic Human and Non-Human Estuarine Ecologies

Heather Green and Owain Jones

12.1 Introduction

Tides, especially ones connected to estuaries and mudflats, perform polyrhythmic writings and erasings; revealing their ghostly traces as the tide ebbs. Each set of tracings, unique, is wiped away by the following high tide, and a new ebb unveils a fresh set of inscriptions. As quintessentially liminal spaces, these landscapes provide a fleeting platform for entanglements with humans and non-humans—and each state of material becoming is haunted by what has recently been, and what is soon to become again. Tides also influence the movements, cultures, and livelihoods of people who inhabit their shores, with the building of harbours,

H. Green (✉)
Arizona State University, Tempe, USA
e-mail: heather.j.green@asu.edu

O. Jones
Bath Spa University, Bath, UK

C. Certomà (ed.), *Blue Kinships*,
https://doi.org/10.1007/978-3-031-78619-8_12

bridges, and embankments along their ever-shifting margins and a long history of regional fishing and folklore.

The creative project *Tidal Timespace: Imprints & Palimpsests* was originally conceived and developed by Heather Green to explore tidal phenomenon, intertidal kin, and tidal humanities in two distinct locations—one in her familiar bay, Bahía Adair in Sonora, Mexico, and another in the distant location of the Severn Estuary, UK (see Green, 2024). She chose the Severn not only because of its impressive tidal range and striking contrasts to Bahía Adair, but also because it is home to geographer Owain Jones, who has written extensively on phenomenological rhythms and cultures of tides. For the Severn Estuary part of Tidal Timespace, Owain, and Heather are collaborating on fieldwork, coauthoring the lexicon, and exploring the Severn's coasts on foot while reflecting on personal geographical loves and losses. When complete, the project will have identical elements for each site, including a series of mudflat casts, a tidal sequence concertina stretched across a metal stand (Picture 12.1), and a series of artist books that interweave local tidal phenomena, regional species, landmarks, histories, and non-human traces to render the granular distinctions of each place.

The ephemeral scripts and traces revealed at low tide read like benthic palimpsests—and for the project Heather has been capturing them in a series of plaster casts that record the still-wet, intricate patterns inscribed by invertebrates, currents, and other beings (Pictures 12.1 and 12.2). Each cast is located on a map, depicting diversity of intertidal patterns, fauna, and loci. The fieldwork, casting and writing for Bahía Adair was completed between 2018 and 2023, and parts of it exhibited in 2023. The fieldwork for the Severn Estuary is being conducted collaboratively with Owain in spring 2024 with support from a Fulbright Scholar grant hosted by Cardiff University and the Severn Estuary Partnership. The Bahía Adair book is now being translated into Spanish, and once the Severn Estuary text is complete and translated into Welsh, both books will be printed and bound, and ultimately exhibited together.

By examining these two landscapes side-by-side, we (Heather and Owain) aim to share both subtle and distinct characteristics as well each other's care of these places. Located in the Upper Gulf of California, Bahía Adair has the third-largest tidal range in North America (Flessa, Fürsich 1991) and is a sparsely populated wetland complex fringed with salt pans, freshwater springs, and multiple esteros (also known as inverse or negative estuaries, they are hypersaline estuaries occurring in dry

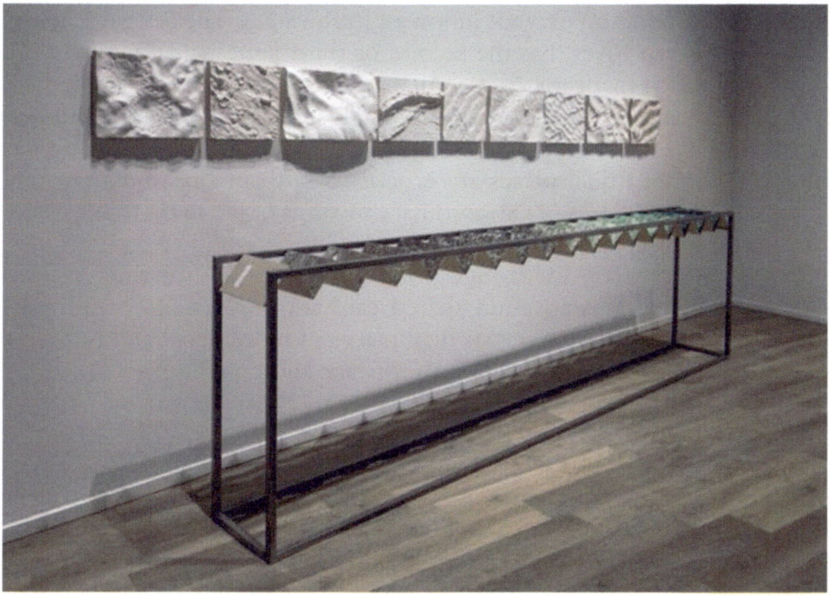

Picture 12.1 Tidal Timespace, installation view, 2023, plaster, steel, inkjet on Asuka. (Photo Heather Green)

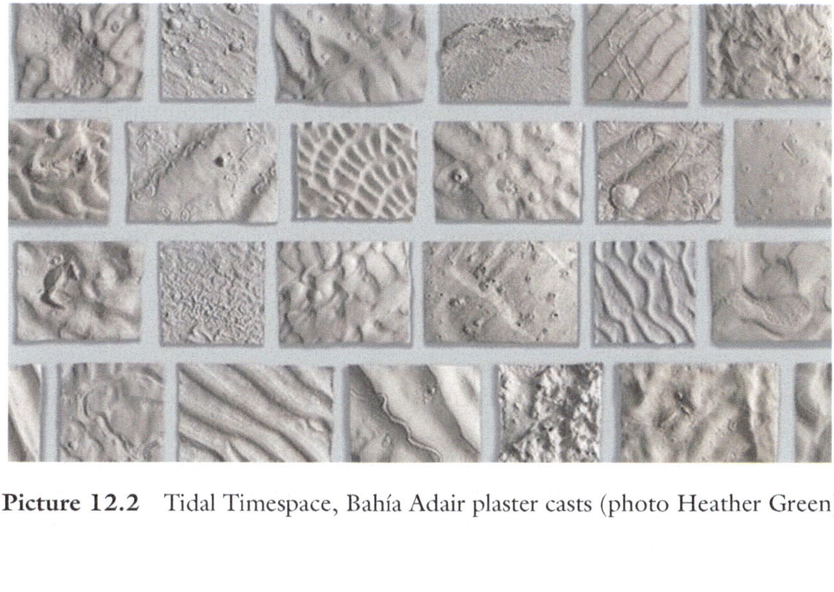

Picture 12.2 Tidal Timespace, Bahía Adair plaster casts (photo Heather Green)

climates as evaporation exceeds inflow of freshwater). The Severn Estuary, fed by the Severn River, has the second or third-largest tidal range in the world, and a rich maritime culture that has been cultivated for millennia. Both places have sustained environmental damage and face continued threats and the risk of climate change-driven transformation. Estuaries, deltas, and related tidal places are crucial to both marine and terrestrial ecology. They are places of extraordinary mixings of life in rhythmic cycles and sites of kinship between many forms of life.

In what follows we discuss questions of coastal and estuarine ecology, marine/terrestrial kinships, and shared care and love of place and non-human life. We describe our two tidal places in more detail and outline a practice of deep mapping for the sake of care and community.

12.2 Kin Networks and Conversive Waters

Haraway's notion of 'staying with the trouble' and 'kin' present challenging ways to rethink land-sea relations. Kinship means 'recognizing, in fact embracing, enduring mutual, obligatory, non-optional, you-can't-just-cast-that-away-when-it-gets-inconvenient, enduring relatedness that carries consequences.' (2019 online) Haraway adds that kinship is bound to be place-specific and uses her own coastal home as an example. 'To be kind is to be kin, but kin are not kind. Kin is often quite the opposite of kind. It's not necessarily to be biologically related but in some consequential way to belong in the same category with each other in such a way that has consequences. If I am kin with the human and more-than-human beings of the Monterey Bay area, then I have accountabilities and obligations and pleasures that are different than if I cared about another place. Nobody can be kin to everything, but our kin networks can be full of attachment sites.' (ibid).

Kinship needs to pay heed to the specificities of places and the specificities of the relational ecologies tracing thorough them. And somehow our kinships with watery places need to extend into the very 'other' space that is the ocean, and the very other types of lives found there (Jones, 2000). Thus, the Tidal Timespace project does this, by paying close attention to sometimes strange, fleeting presences and traces of life in the intertidal zones of 'our tidal places.'

As we will show in more detail, one of the extraordinary things about intertidal land is that it is 'rewritten' after each tide. And in terms of kinship, it a co-dwelt space, being sea for some hours, and then land again. It thus joins sea to land and does so in ways which are ecologically vital. Estuaries, by definition, have a landward ingress of freshwater, and tide-driven seaward ingress of saltwater, often with large areas of intertidal land exposed at low tide. They are thus spaces where the most profound divisions of planetary material space, between salt and freshwater, and between land and sea, are subject to ceaseless rhythmic flux through the complex ebbing and flowing rhythms of the tides, in cycles which overlay daily, monthly, and seasonal patterns. Again, this is a vital zone for global ecology.

The earth's biosphere is expression of kinship between watery and terrestrial becomings. In the book *Wildwood* Roger Deakin makes the point that rivers connect the ecologies of woods and forests to ecologies of the oceans. Rivers carry organic matter such as leaf and wood litter and discharge it into the sea where it is an important foundation of ocean ecology, for example. Colin Welland (2009) similarly considers that rivers are the 'arteries' of the earth as they transport sediment and organic material from evolving land to evolving oceans over eons. This is expressive of the idea of the earth as a living body of processes which modernity has been so dismissive of. Welland goes on to point out that 'around a third of this natural volume is [now] prevented from reaching the oceans as a result of being trapped behind dams and other man-made obstructions' (81).

The word 'estuary' is derived from the Latin *aestus* meaning tide, referring to a tongue of the sea reaching inland (Hardisty, 2008); a rhythmic conversation and exchange between two bodies. But as well as a 'tongue of the sea' reaching inland on tidal rivers, all river mouths extend a tongue, a plume of fresh water into the sea. Smith (2002) in his study of the extraordinarily rich ecologies and cultures of the Amazon River mouth, retells reports from early European explorers, who, while still sailing far off the coast in Atlantic waters, knew they were on course for the mouth of some huge river by realizing they were sailing in streams of differing-coloured waters.

Where the coast is formed by the lower reaches of wetlands or rivers, particularly those that become estuaries, such as the mighty Amazon, saltwater flows up through the mouth as the tide rises, and out again as the tide ebbs. Freshwater meets saltwater, and these mix to make brackish water. But even in the case of inverse estuaries as with Bahía Adair, tidal inflow reaches up and blends with the saltier water in the estuary's channels, creating a dynamic habitat that functions as a nursery for many marine species and sanctuary for migrating birds. Flora and fauna, organic and inorganic matter, and human movement all pass through estuarine mouths, and distinctive ecologies and cultures form in these unique places. A disproportionate amount of global biodiversity is conceived and sustained in coastal zones, including deltas, estuaries, and liminal margins such as these.

Tides affect all seas and coasts, to varying degrees (Carson, 1951), and the ways in which people interact with them (Aldersey-Williams, 2016). This 'greatest force on earth' (ibid, cover) affects estuaries most significantly of all. They are 'ecotones' as defined by Neimanis (2012); places of 'watery relationality,' highlighting the interconnected and mutually constitutive social, cultural, and environmental relationships between bodies of water and human and non-human bodies (Denning, 2021). Despite the very high ecological and cultural richness of these places, they remain poorly explored across the social sciences and the arts and humanities, and they seem politically and culturally under-regarded given their extent, importance, and the existential risks they now face.

Jung et al.'s, recent study of global estuaries (2024), have the distressing title, *Economic Development Drives Massive Global Estuarine Loss in the Anthropocene*. In it they describe how estuaries around the world have been diminished and degraded by various developments and pressures. Some have been entirely erased. To this slow progression of destruction, over recent centuries, is now added the severe risks of climate change. These involve the obvious factor of sea level rise, and also changes to sea salinity, erosion via storm surges and increased wave action, and changes to freshwater flows. In geological time, estuaries have migrated up and down river valley channels over eons, responding to changing sea levels. The big problem now is that highly engineered and developed estuaries, with flood protection and controlled river channels, render estuarine and wetland migration inland nearly impossible, or highly problematic for settlements. A number of world cities such as London and New York are built around estuaries mouths, which were previously some of the most biodiverse landscapes on earth.

12.3 Attention, Attachment and Care

Bahía Adair is a large embayment in the northern Gulf of California in Sonora, Mexico, and is the native home of the Hia C-eḍ O'odham people, the southern border of the Cucupá's land, and until as recent as the

1950s, a pilgrimage destination for the Tohono O'odham. When looking at a map, especially when looking at a satellite photo, the geographic features bewilder with large areas of esteros, serpentine tidal channels, and virgin coasts painted in contrasts of pale sands and ultramarine blues and greens. Positioned in a remote area of the Sonoran Desert with no natural fresh water source aside from a scattering of artesian springs, and with no infrastructure until quite recently, the area has been slow to develop and is still relatively pristine. When spring tides ebb the water goes far out beyond the horizon. Mudflats cover much of the intertidal area and are firm underfoot due to the bioturbation from stingrays and ghost shrimp—and their patterns are diverse, inscribed by many animals and sculpted by the agency of water and tidal forces.

A city just south of Bahía Adair, Puerto Peñasco was established in the 1930s when fishermen were drawn to the area for the abundant sea life in the Northern Gulf of California. Boat builders, a railroad line, and shipped-in water followed. Now the city hosts a bustling population of around 70,000. It is close to the Mexico/US border, and as a result, has become a popular tourist destination for US citizens—many of whom own vacation homes in the area, including Heather's family—her grandfather built a small beach cottage in the adjacent village of La Choya in the early 1960s.

Located on a small peninsula at the southern tip of Bahía Adair, La Choya didn't have any street names until the 1990s, and even beaches weren't given proper local names, but named instead by visiting American tourists—Tucson Beach, for example. The rocky promontory on the peninsula is listed with a different name on every map! It has been named Roca del Toro, Sea Lion Bluff, Punta La Choya, and US tourists call it simply 'Competition Hill' for the many noisy and destructive dune buggies that race up its one steep sandy slope. There are no museums or community libraries in La Choya, so unless you look online, visit the archives at CEDO Intercultural, or search within a distant university library, much of its natural and cultural history is sequestered. When the peninsula was incorporated into Sandy Beach Resorts, developers did name the streets, but residential streets were named simply with letters of the alphabet. Heather's family's cottage is on Calle N. Unnamed areas are free for the taking, it would seem. As such, she has been keen to uncover place names, species names, and the deeper history of La Choya, and has explored archival themes of enumeration in much of her creative research. 'Untold land is unknown land,' states Lucy Lippard (1997)—names, maps, and photographs can authenticate our communal imagination, history, and identity, and often validate stewardship and conservation.

Tidal Timespace gave Heather the opportunity to explore the greater reaches of Bahía Adair and better understand La Choya's place within the larger geographic whole. René Loaíza from CEDO Intercultural helped her find a local fisherman to interview, Rafael Peñuelas Machuca—who agreed to write his incredible story about growing up in Peñasco fishing with his father in Bahía Adair. René also put her in touch with diver Ernesto Gastellum whom she hired to assist in the field making casts of the mudflats, and who ultimately became her colleague travelling out into the far reaches of Bahía Adair by boat or four-wheel drive. The coastal and adjacent areas along Bahía Adair north of La Choya are part of local *ejidos*, or government-granted land, so essentially undeveloped, private lands. And even with a recent coastal highway along its fringes, the larger bay and wetlands remain relatively inaccessible and mysterious. Much of the time it was a first for both Ernesto and Heather to visit these places—so the uncertainty created a sense of adventure.

The Severn Estuary is the second largest estuary in the world, something like 556 sq km, roughly 18% of which is intertidal (100 sq km). The Estuary is formed from the lower reaches of the River Severn, just above the City of Gloucester and the upper Bristol Channel, which are highly tidal. The River Severn is the longest river in Great Britain with the most voluminous outflow of fresh water by far in of all the rivers of England and Wales, into the Bristol Channel, which runs on west to join the Celtic Sea and Atlantic Ocean. The estuary has a shoreline of about 353 km, 80% of which is lined with sea walls. These are mostly grassed earthen banks in rural locations and concrete walls in and around urban areas.

The Estuary has a rich maritime culture, from vernacular sailing craft (trows and pilot cutters) to regional fishing techniques (lathe nets, putchards), which were shaped to deal with the enormous tides. The Severn Estuary is 'macro-tidal,' having some of the highest tides in the world. The 'springs' rise about 14.5 metres (at Avonmouth) in the space of 7 hours, only to fall back a similar height in the following ebb. This elementally profound, twice-daily physical transformation of the landscape has endless implications for nature, culture, and economy and creates what we call 'tidal cultures.' The higher tides create the Severn Bore, a celebrated natural phenomenon, which is a large tidal wave that washes up the tidal stretches of the river channel.

About one million people live around its shores, making it the most populated estuary with mega-tides in the world. There are large ports, nuclear and other power stations, two huge road bridges and a rail tunnel crossing at mid estuary. There are also seaside resorts in the lower reaches on both the English and Welsh shores. Between the towns and cities, rural areas exist with small villages, many of which were once small harbours. Despite the developments, the estuary is a hugely wild place with very rich and important natural history, particularly wading and water birds. The ecological and landscape richness of this estuary has already been reduced over many centuries by the building of seawalls and river engineering which have turned coastal marshlands to farm and development land. But the sea walls are, in many places, openly accessible and used for agricultural and recreational purposes (e.g., a long-distance footpath; The Severn Way) and accessing and witnessing the intertidal areas, the ever-changing views, and the natural wonders and cultural curios that remain. As in Bahía Adair the tides are the central feature of this place. Driven by the relational movements of the earth, moon, and sun, they have varying daily monthly and seasonal rhythmic cycles which over lay the more ubiquitous cycles of night and day and the seasons in complex ways.

12.4 Human and Non-Human Verse

Many things about the sea bear resemblance to texts: cursive scripts made by snails; curling waves turning as pages upon sand; stratified volumes encased in sea cliffs. The countless textures left at low tide uphold patterns of tidal rhythms and cast wavelets that have been reworked as a palimpsest, and lay ground for what will soon be erased and rewritten again. Tidal landscapes enact more-than-human rhythms, attuning to them helps us remember and imagine the possibility of a different kind of livability.

Lexicographical research—the art of compiling dictionaries, or the act of collecting all the words within a single subject, language, or dictionary—provokes questions about the validity of categories and binaries like nature/culture, human/non-human, and English/Spanish. Books can provide orientation, intrigue, and wayfinding for the reader while the act of reading creates possibilities for the book as a site of imagination and discourse. The hope with this project is that our book can become a vessel for multiple viewpoints and can extend beyond a physical exhibition and continue life beyond the four walls of a gallery.

We seek to represent humans/non-humans, Spanish/English, and Welsh/English with equal weight—so plants and animals are addressed with personhood and distinction, and given proper pronouns (in English at least), and each language can be read individually rather than as a hierarchical translation. The unique A-Z order of the Spanish and Welsh lexicons as a separate text from that of our native English language lexicon allows for imagining the quotidian familiarity of common names in another language and culture.

In the centre of the Bahía Adair book a third 'lexicon' is written by Pygmy Olive Snails. These small marine snails (*Olivella dama*), about ¾ of an inch long, inhabit the sandy intertidal substrate of the rocky cove on the south end of Bahía Adair, below Heather's family's cottage. When the tide recedes, they burrow under the sand, concealing their highly polished and ornate shells. Only visible out of water during the lowest spring tides, they appear as raised bumps under the smooth and damp virgin sand, hiding to seek shelter. Then they begin their slow cursive travel as they search for food, or possibly each other, twisting and turning as they sense out their destinations. These movements create beautiful and often complex designs, at times accidentally resembling recognizable letters of the Roman alphabet. It started with just a few letters…H, e, f, Q, but soon became a kind of mission to find a complete set. There were a few photographs that needed to be flipped or rotated to make legible—but otherwise they are unaltered (Picture 12.3). You'll see however, the Olive's native 'language,' the scripts that aren't recognizable as human symbols, are far more arresting to look at. There's a wonderful German word that field philosopher and writer Thom Van Dooren introduces in his book *A World in a Shell*, 'streifzuge'— describing the forays or imaginative excursions into another species' umwelt. We can only dream what these snails might be up to as they create their intricate traces at low tide—but imagining their lives is an effort worthwhile!

Interspersed within the lexicon are a variety of vignettes, fragmented memories that break up the entries and help contextualize them. There are also invited comments and marginalia from a variety of people—scientists, fishermen, long-time residents, and artists—with the intention of adding more interest, perspectives, and a sense of community. This process of gathering and researching these many names, terms, and place-words, along with the other multi-media elements in the larger installation, are embedded in the practice of deep mapping—a methodology that intertwines ecological and historical narratives, personal and

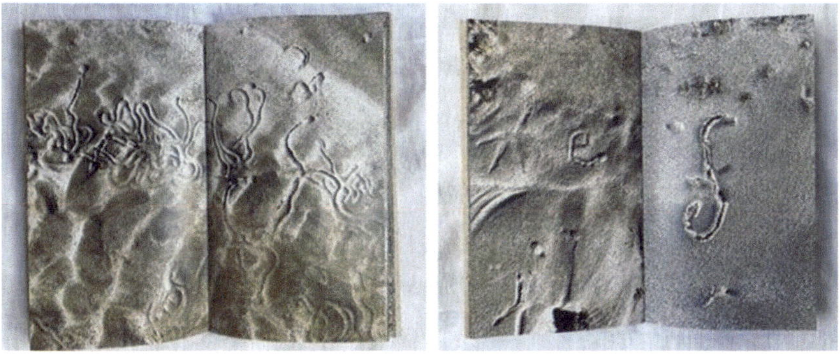

Picture 12.3 Tidal Timespace, Pygmy Olive Scripts, inkjet (photo Heather Green)

communal memory, scientific data, and a wide range of media into a rich, a multivocal project. McLucas (2010) describes how deep mapping 'brings together the amateur and the professional, the artist and the scientist, the official and the unofficial, the national and the local.' The use of deep mapping as an aesthetic and methodological instrument democratizes knowledge by weaving temporal, spatial, and disciplinary threads into one practice (Springett, 2015).

'The more you know, the more you know you don't know,' as they say, and even if there was more time, it would be impossible to list every plant, animal, person, or place name, however small the place may be, which is not the point of the project. There wasn't enough time to gain the trust and get to know anyone living on the ejidos, something that is lacking in the Bahía Adair collection of place-based names and lore. And there are no citations—as this is not a scientific or academic text *per se*, nor a traditional regional or geographical text—but a creative practice of care that seeks to learn about and share knowledge, share the thrill of exploring in the field, and evoke a fineness of texture that makes these places unique. For us it's also an invitation to collaborate with others and ask for their perspectives and stories.

At the time of writing this, we are in the midst of coauthoring the Severn Estuary lexicon and exploring the coast on foot while reflecting on our personal geographical loves and losses. Owain grew up on a farm on the Welsh side of the estuary on the Wentlooge Level, and when he was in his early 20 s one area of coastal saltings used for the grazing of livestock was lost to development for landfill, and the rest of the farm taken by the city council for housing though a compulsory purchase process. And in the 1990s around 80% of Estero La Choya was destroyed to create a golf course and residential resort—so we share a grief of lost childhood landscapes. But we also share a propensity to continually renew our kinship with these places, returning again and again, discovering new things in the process.

Juxtaposing two distinct environments provides a structure for careful study and a unique method of acknowledgement that goes far beyond a binary act of comparison or 'discovery' in the classic sense. Although much of this collaborative project has been accomplished at home or through long-distance correspondence, the spirit is performed in the field—images and ideas string together in walking, listening, tracing, conversing, reflecting, and collecting. As Les Roberts (2016) states, 'The "map" is lodged in the more immaterial spaces of the body and imagination. Its performativity is made flesh in the way the walker inhabits and dwells within the space that both map and walker conjure into being.' Timing our fieldwork on the flats during low tide, we submit to chance encounters and discoveries, to weather, and to spaces of silence and exchange. The nomenclature we collect echoes the meander of walks and conversations, so that our cataloguing becomes an act of enchantment and placemaking in contrast to a stratified collection system—positioning the past in relation to the present and employing memory to transcend colonial archival thinking (Zaayman, 2023).

12.5 Conclusions

It would be an impoverished sensibility if one did not feel, when on intertidal land at low tide, the strangeness of the same space at high tide. This is part of the wonder of estuaries and other tidal places and part of their vulnerability. And all this is overlain now by a darker future spectre of climate change, sea level rise and storm surge induced erosion. Just as the mudflats become a palimpsest after the tide's daily erasing of the cursive travel of snails and other scripted impressions, so too, do many man-made pursuits; the overwriting for development obliterates rich ecological stories and effaces local narratives.

These landscapes currently face many anthropocentric threats, some already erased, some haunted as imagined futures. In Bahía Adair, rising sea temperatures are bleaching sponges and anemones, unregulated fishing practices are causing species decline (from intertidal invertebrates, reef fish, to the critically endangered endemic porpoise, the Vaquita), in populated areas, condos, golf courses, and off-road vehicles have greatly eroded or destroyed critical estuaries and intertidal habitat, and now there is a threat of a desalinization plant being built. The Severn Estuary has long been overshadowed by the prospect of radical reduction and even erasure by the building of a barrage. The smaller Taff Estuary within the wider Severn Estuary, in Cardiff, has undergone this fate. We hope that by aligning these two tidescapes, the project can create an intriguing contrast and framework for bringing about awareness of climate impacts on estuaries world-wide, which look to be quite severe (Nienhuis et al., 2020).

As the impacts of modern consumption and globalized capitalism on the environment are toxic (Jones, 2024) to the point that we are now in the 6th great extinction event—the era of ecocide, perhaps love is the one strand of 'true' ecology that modern humans still have capacity for. Love for place, for community, and for each other, and maybe most importantly in ecological terms, love of others' love. If love is about deep levels of interconnection, interbecoming, can these stories of shared care of place become part of a wider process of ecological healing?

In the great meta-narratives of 'the land' and 'the ocean' and the politics, cultures and policies thereof, coasts can often be forgotten. The coasts of the world are usually considered as edges—of both the land and the sea. But seen in a different way, coasts can be imagined as the centres of things. They are critical thresholds for life and culture in many ways, going against the usual notion of borders as 'unfixed, fluctuant, and infinitely permeable'... 'An open border but a dangerous one between the known and the unknown, which only a few sibyls, amphibians, crustaceans and marine mammals traverse with impunity.' (Solnit, 2007). So many cultural and ecological tracings pass through the coast, and particularly through the thresholds that are estuaries, deltas, and river mouths—where the salt and the fresh mingle— and in those passings, are transformed and feed their agency into our planetary becoming. Here are some of the richest ecotones that exist. They teem with life, both in place, and life passing through.

We hope the attention to, and care of, those traces of that life the project has created brings the life of the tidal coast to wider awareness. We have shown how such places have been under a range of pressures as modern society continues to voraciously develop and despoil, not least along the world's coasts. To those pressures is now added the challenge of climate change. Estuaries are particularly at risk. So the ecological webs, the kinships, of these places as also at risk. This causes us, and others, much concern.

Acknowledgements Both the Bahía Adair and Severn Estuary parts of this project have been helped by contributions from many other people. We thank all. We also thank the editors for their support and constructive comments. Heather's time in the UK was supported by a Fulbright Scholar Award hosted by the School of Earth and Environmental Sciences at Cardiff University and the Severn Estuary Partnership, and her home institution Arizona State University.

References

Aldersey-Williams, H. (2016) Tides. The Science and Law of the Greatest Force on Earth, London: Penguin Random.

Carson, R. (1951). *The sea around us*. Oxford University Press.

Deakin, R. (2007). *Wildwood*. Hamish Hamilton.

Denning, L. (2021). Ecotone as methodology. *GeoHumanities*, *7*(1), 1–12.

Denning, L. (2023) Interim Bangalore: Bodies as sensory data collectors. In A. Pigott, O. Jones, & B. Parry, (Eds.), *Art and creativity in an era of ecocide: Embodiment, performance and practice*. Bloomsbury.

Flessa, K., & Fürsich, F. (1991). Ecology, Taphonomy, and paleoecology of recent and Pleistocene molluscan Faunas of Bahia La Choya, Northern Gulf of California. *Zitteliana*, *18*, 5–180.

Green, H. (2024). Tidal timespace. https://www.heathergreen-art.com/tidal-timespace.

Hardisty, J. (2008). *Estuaries: Monitoring and modelling the physical system*. Blackwell.

Haraway, D. (2019). *(In conversation with Steve Paulson); Making Kin: An Interview with Donna Haraway*. Los Angeles Review of Books, 09 12 2019, https://lareviewofbooks.org/article/making-kin-an-interview-with-donna-haraway/ Accessed 16 06 2024.

Jones, O. (2000). Inhuman geographies: (un)Ethical spaces of human non-human relations. In C. Philo & C. Wilbert (Eds.), *Animal spaces, beastly places: New geographies of human-animal relations* (pp. 268–291). Routledge.

Jones, O. (2024). Toxic Dwelling? Speculations on why transformations away from the ecocidal impacts of modern globalised consumer cultures of capitalism are proving so hard to make. *Hyphen Journal*, forthcoming. https://hy-phen.space/journal/

Jung, N. W., Lee, G., Dellapenna T. M., Jung, Y., Jo, T., Chang, J., & Figueroa S. M. (2024). Economic development drives massive global estuarine loss in the Anthropocene. *Earth's Future, 12*, 4.

Lippard, L. (1997). *The lure of the local: Senses of place in a multicentered society* (p. 46). The New Press.

McLucas, C. (2010). There are ten things that I can say about these deep maps. https://metamedia.stanford.edu/~mshanks/projects/deep-mapping.html. Accessed (by Biggs, I.) 28 March 2010.

Neimanis, A. (2012). Hydrofeminism: Or, on becoming a body of water. In H. Gunkel, C. Nigianni, & F. Söderbäck (Eds.), *Undutiful daughters: Mobilizing future concepts, bodies and subjectivities in feminist thought and practice.* Palgrave Macmillan.

Nienhuis, J. H., Ashton, A. D., Edmonds, D. A., Hoitink, A. J. F., Kettner, A. J., Rowland, J. C., & Törnqvist, T. E. (2020). Global-scale human impact on delta morphology has led to net land area gain. *Nature, 577*(7791), 514–518.

Pearson, M. (2006). *In comes I: Performance, memory and landscape.* University of Exeter Press.

Roberts, L. (2016) Deep Mapping, Basel: MDPI.

Solnit, R. (2007). *Storming the gates of paradise: landscapes for politics.* University of California Press.

Springett, S. (2015). Going deeper or flatter: Connecting deep mapping, flat ontologies and the democratizing of knowledge. *Humanities, 4*, 623–636; https://doi.org/10.3390/h4040623

Smith, N. (2002). *Amazon sweet sea: Land, life, and water at the river's mouth.* University of Texas Press.

Van Dooren, T. (2021). *A world in a shell: Snail stories for a time of extinctions.* The MIT Press.

Welland, M. (2009). *Sand. A journey through science and imagination.* Oxford University Press.

Zaayman, C. (2023). *Anarchival practices: The Clanwilliam arts project as re-imagining custodianship of the past.* ICI Press.

CHAPTER 13

Blue Kinship. Documenting Ocean Relationships

Federico Fornaro and Giuseppe Lupinacci

[The video reportage is available at the link https://youtu.be/Ej0ICNcfPkU, access is allower via the following system https://www.raw-news.net/blue-kinship-documenting-ocean-relationships/].

Supplementary Information The online version contains supplementary material available at https://doi.org/10.1007/978-3-031-78619-8_13. The videos can be accessed individually by clicking the DOI link in the accompanying figure caption or by scanning this link with the SN More Media App.

F. Fornaro (✉) · G. Lupinacci
Raw-News Visual Production Agency, Epping, UK
e-mail: federico.fornaro@raw-news.net

C. Certomà (ed.), *Blue Kinships*,
https://doi.org/10.1007/978-3-031-78619-8_13

Index

C. Certomà (ed.), *Blue Kinships*,
https://doi.org/10.1007/978-3-031-78619-8

The manufacturer's authorised representative in the EU is Springer Nature Customer Service Centre GmbH, Europaplatz 3, 69115 Heidelberg, Germany. If you have any concerns regarding our products, please contact ProductSafety@springernature.com

Printed and bound by CPI Group (UK) Ltd, Croydon, CR0 4YY
30/09/2026
02225507-0002